Springer Geography

The Springer Geography series seeks to publish a broad portfolio of scientific books, aiming at researchers, students, and everyone interested in geographical research. The series includes peer-reviewed monographs, edited volumes, textbooks, and conference proceedings. It covers the major topics in geography and geographical sciences including, but not limited to; Economic Geography, Landscape and Urban Planning, Urban Geography, Physical Geography and Environmental Geography.

Springer Geography—now indexed in Scopus

More information about this series at http://www.springer.com/series/10180

Wei Zhou · Jianlong Li · Tianxiang Yue

Remote Sensing Monitoring and Evaluation of Degraded Grassland in China

Accounting of Grassland Carbon Source and Carbon Sink

Springer

Wei Zhou
Department of Geographic Information
and Land Resources
Chongqing Jiaotong University
Chongqing, China

School of Geographical Sciences
Southwest University
Chongqing, China

Tianxiang Yue
Institute of Geographic Sciences and Natural
Resources Research (IGSNRR)
Chinese Academy of Sciences
Beijing, China

Jianlong Li
School of Life Science
Nanjing University
Nanjing, China

ISSN 2194-315X   ISSN 2194-3168 (electronic)
Springer Geography
ISBN 978-981-32-9381-6   ISBN 978-981-32-9382-3   (eBook)
https://doi.org/10.1007/978-981-32-9382-3

This Springer imprint is published by the registered company Springer Nature Singapore Pte Ltd.
The registered company address is: 152 Beach Road, #21-01/04 Gateway East, Singapore 189721,
Singapore

Contents

1 **Research Progress of the Grassland Carbon Cycle and Grassland Degradation in China** 1
  1.1 General Situation of Grassland in China. 1
  1.2 Related Definitions and Model Formulation 4
    1.2.1 Basic Process of Carbon Cycle 4
    1.2.2 Observation and Simulation of NPP and Carbon Sinks in Grassland Ecosystem. 5
  1.3 Literature Review 7
    1.3.1 Overview of Grassland Degradation 7
    1.3.2 Grassland Degradation Monitoring Method 8
    1.3.3 Driving Factors of Grassland Degradation 9
    1.3.4 Ecological Restoration Project. 11
  1.4 Problems in Carbon Sinks Accounting and Grassland Degradation Research 12
    1.4.1 Studies on the Driving Mechanism of Grassland NPP in China Are Few 12
    1.4.2 Research on the Grassland Carbon Sinks Evaluation in China's National Scale Area Few 12
    1.4.3 Lack of Remote Sensing Monitoring and Driving Mechanism Quantitative Assessment of Grassland Degradation in China 13
  References 13

2 **Grassland Coverage Inter-Annual Variations and Its Coupling Relation with Hydrothermal Factors in China** 17
  2.1 Introduction 18
  2.2 Methodology 19
    2.2.1 Normalized Difference Vegetation Indices (NDVI) 19
    2.2.2 Meteorological Data. 19
    2.2.3 Land-Cover Products 19

      2.2.4   Estimation of Grass Coverage and Its Accuracy
             Validation   20
      2.2.5   Grassland Vegetation Coverage Trend Analysis   22
      2.2.6   Correlation Analysis of Grass Coverage and Climate
             Factors   23
  2.3   Implementations and Comparisons   23
      2.3.1   Spatial Distribution of Long-Term Mean Grass
             Coverage   23
      2.3.2   Temporal Changes of Grass Coverage in China   24
      2.3.3   Correlation Analysis of Grass Coverage and
             Temperature, Precipitation   27
      2.3.4   Comparisons   32
  2.4   Conclusions   35
  References   36

**3 Spatial–Temporal Dynamics of Grassland Net Primary
  Productivity in China and Its Response to Climate Factors**   39
  3.1   Introduction   40
  3.2   Data Source and Processing   41
      3.2.1   GIMMS NDVI Data and Processing   41
      3.2.2   Meteorological Data and Processing   41
      3.2.3   Grassland-Type Data   41
  3.3   Methodology   42
      3.3.1   CASA Model   42
      3.3.2   Validation of CASA Model   43
      3.3.3   Inter-annual Variation Trend of Grassland NPP   44
      3.3.4   Correlation Coefficient   45
  3.4   Results and Discussions   45
      3.4.1   Spatial Distribution Characteristics of Grassland NPP
             in China   45
      3.4.2   Time Dynamic Analysis of Grassland NPP in China   46
      3.4.3   Correlation Analysis Between Grassland NPP and
             Temperature and Precipitation   48
      3.4.4   Discussions   51
  3.5   Conclusions   52
  References   54

**4 Inter-annual Variation in Grassland Net Ecosystem Productivity
  and Its Coupling Relation to Climatic Factors in China**   55
  4.1   Introduction   56
  4.2   Data Sources   57
      4.2.1   Study Area   57
      4.2.2   Data Source and Processing   59

4.3 Methodology 60
    4.3.1 BEPS Model Description 60
    4.3.2 Research Indicators and Statistical Analysis 61
    4.3.3 Implementations and Discussions 63
    4.3.4 Spatial Distribution of Grassland NEP in China 64
    4.3.5 Inter-annual and Monthly Changing Trend
           of Grassland NEP in China..................... 66
    4.3.6 Inter-annual Variation Trend of Grasslands'
           CSE and RUE in China 67
    4.3.7 Correlation Analysis of Grassland NEP with
           Temperature and Precipitation................... 68
    4.3.8 Lag Analysis Between Grassland NEP and Precipitation
           and Temperature 70
    4.3.9 Connection Between PD and Grassland NEP.......... 71
    4.3.10 Comparisons 72
4.4 Conclusions 77
References 78

5 The Variation of Landscape and NPP of Main Pastoral
   Grasslands in China 83
   5.1 Introduction 83
   5.2 Methodology 85
       5.2.1 Study Area 85
       5.2.2 Land-Use Data............................. 85
       5.2.3 Landscape Metrics 85
       5.2.4 CASA Model to Estimate NPP 86
       5.2.5 Model Accuracy Verification 87
   5.3 Implementations and Discussions........................ 88
       5.3.1 Analysis of Spatial and Temporal Dynamic Changes
              of Grassland 88
       5.3.2 Changes in Pattern of Grassland Landscapes 90
       5.3.3 Changes in NPP of Grassland................... 93
       5.3.4 Comparisons 95
       5.3.5 Effects of Ecological Restoration Measures
              on Grassland Productivity...................... 100
   5.4 Conclusions 102
   References 102

6 Grassland Degradation Remote Sensing Monitoring and Driving
   Factors Quantitative Assessment in China from 1982 to 2010 105
   6.1 Introduction 105
   6.2 Methodology 107
       6.2.1 Study Areas............................... 107
       6.2.2 Data Sources and Processing 108

6.2.3   Model   109
6.2.4   Calculation of Fractional Vegetation Cover   111
6.2.5   Slope of Grassland NPP   112
6.2.6   Grassland Degradation Status Evaluation   112
6.2.7   Scenarios Design and Quantitative Assessment Method
        of Grassland Degradation   113
6.3   Implementations and Discussions   113
6.3.1   Spatial Distribution of Grassland Degradation Status   113
6.3.2   Quantitative Assessment of Grassland Degradation
        or Restoration in China from 1982 to 2010   115
6.3.3   Comparative Analysis of the Driving Contribution
        of Two Factors in the Nine Provinces   117
6.3.4   Discussion   117
6.4   Conclusions   120
References   120

7   **Grassland Degradation Restoration and Constructing Green
    Ecological Protective Screen**   125
7.1   Overview of the Green Ecological Protective Screen Policy
      in China   125
7.2   Restoration Technology of Degraded Grassland   126
7.2.1   Grassland Degradation in China   126
7.2.2   Restoration of Degraded Grassland   128
7.2.3   Restoration Technology of Degraded Grassland   128
7.3   Sustainable Use of Rangeland Resource   130
7.3.1   Grassland Improvement and Artificial Grass Planting
        in Pasturing Areas   130
7.3.2   Strengthening Laws and Regulations and Implementing
        Grassland Law   131
7.3.3   Control the Excessive Utilization of Grassland   131
7.3.4   Improving the Compensation Mechanism for Grassland
        Ecological Construction and Strengthening the Reward
        Mechanism for Ecological Protection   132
7.3.5   Establishing a Comprehensive System of Grassland
        and Livestock Products Industrialization According
        to Local Conditions   133
7.3.6   Adjusting Ecological Restoration Project Reasonably   134
7.4   Construction of Grassland Green Ecological Protective Screen . . .   135
References   137

Chapter 1
Research Progress of the Grassland Carbon Cycle and Grassland Degradation in China

Abstract Grasslands, one of the most common vegetation types in the world, account for nearly 20% of the global land surface. The vast land cover and carbon sequestration potential make it become the important composition of the terrestrial carbon cycle. In China, grasslands mainly locate in the arid and semi-arid areas in the northwest and the Tibetan Plateau alpine climate regions, which make it sensitive and vulnerable to climate change and frequent human intervention. Along with the global warming and population explosion, substantial land-use and cover change has occurred in China due to overgrazing, grassland reclamation, and over-mining explorations. All these changes had led to serious ecological problems, such as grassland degradation, desertification, and future changes of the global carbon cycle. In recent decades, Chinese government has initiated several ecosystem restoration programs to mitigate the environment degradation. Meanwhile, along with the development of remote sensing technology, grassland carbon sequestration can be simulated by ecological remote sensing model in large scale, and previous studies showed that grassland of China is carbon sink. Grazing is one of the main ways of grassland resource utilization and is also the main factor of grassland degradation. About 35% of the degraded grasslands in the world are caused by overgrazing, which accounts for more than 20% in China. However, the spatial–temporal dynamic of carbon sink or source of grassland of China, and the driving mechanism of grassland degradation in China need deep analysis.

Keywords Grassland ecosystem · Carbon sink · Grassland degradation · Ecological project · Remote sensing

1.1 General Situation of Grassland in China

Grassland is one of the most widely distributed vegetation types on the earth's surface. The global grassland area is about 24 million km^2, accounting for 20% of the global land area. Alpine, tropical, and temperate natural grasslands are very sensitive to regional climate change (Asner and Martin 2004; Wang et al. 2002). At the same time, the huge distribution area may make the grassland play an important carbon sink function in the global carbon cycle (Parton et al. 1993; Scurlock and Hall 1998).

© Springer Nature Singapore Pte Ltd. 2020
W. Zhou et al., *Remote Sensing Monitoring and Evaluation of Degraded Grassland in China*, Springer Geography,
https://doi.org/10.1007/978-981-32-9382-3_1

The study also showed that grassland ecosystem vegetation and soil carbon storage accounted for 34–37% of global terrestrial carbon storage (2500 Pg C) (Matthews et al. 2000). In addition, 90% of the carbon storage of grassland ecosystem was stored in the soil, while the above-ground vegetation biomass accounted for only 10% (Sharrow and Ismail 2004). The rate of soil carbon decomposition in grassland ecosystems is relatively slow, so it is a potential carbon sink (Conant et al. 2001).

Grassland of China is a natural barrier for ecological environment protection. It is mainly distributed in the temperate continental semi-arid climate zone and the Qinghai–Tibet Plateau alpine climate zone. The area is about 4 million km^2, accounting for 41.7% of the country's land area (Ren et al. 2008), accounting for 6–8% of the global grassland area, while China's grassland carbon reserves account for about 9–16% of global grassland carbon storage (Ni 2002). However, the ecological environment of grassland distribution areas in China is fragile and sensitive to climate change. In recent years, disturbances in human activities have intensified, especially overgrazing, grassland reclamation, excavation of herbs, and mining. The combination of warming and drying of the climate and human disturbances led to changes in the structure and function of grassland ecosystems in China, and large-scale degradation and desertification of grassland.

Grassland degradation in China began in the 1960s. In the mid-1970s, the country's degraded grassland accounted for 15% of the total grassland area. With the sharp increase of population in the 1980s, with the sharp increase in population, the demand for food and livestock products increased, grassland overloaded grazing, grassland reclamation and grassland degradation. The phenomenon is becoming more and more serious. By the mid-1980s, the grassland degradation area accounted for 30% of the total grassland area, and in the mid-1990s, it reached more than 50%. The grassland ecological environment is deteriorating (Fan et al. 2007). Grassland degradation will lead to simplification of grassland community structure, degradation of grassland ecological environment function, showing decline in grassland cover, height and productivity, reduction of grassland species diversity, reduction of ecological service value (Levy et al. 2004; Li 1997; Turner II et al. 2001), reduction of perennial herb, and proportion of inedible grass increase (Milchunas and Lauenroth 1993; Milton et al. 1994). At the same time, grassland overload grazing reduces the amount of surface litter and biomass entering the soil and reduces soil carbon input. On the other hand, trampling of livestock also lead to the change of soil water permeability and soil hydro-thermal conditions, increasing soil erosion, eventually leading to a decrease in grassland NPP (Millington et al. 2007; Snyman and Fouché 1991) and an increase in soil organic carbon release (Wang et al. 2002). Severe grassland degradation will lead to a decline in grassland carbon sequestration and a reduction in ecological function.

Climate warming has an important impact on the global carbon cycle and terrestrial carbon balance, and especially in the late 1980s, with the acceleration of global industrialization and urbanization and population expansion, the dynamic changes of the global carbon cycle are closely related to climate change and human activities, and have become the focus of attention of scientists and governments (Houghton et al. 2001). As an important part of the global carbon cycle, terrestrial ecosystems are also important sources and sinks of CO_2 in the atmosphere. Vegetation and soil, as two major components of terrestrial ecosystems, play an important role in the global carbon cycle. Vegetation fixes CO_2 in the atmosphere through photosynthesis, while the respiration of vegetation and soil releases CO_2 into the atmosphere.

Land-use and cover change (LUCC) is one of the ways in which humans can change the natural environment on land. Studies on terrestrial ecosystems show that human activities contribute to one-third of the world's land surface cover as an unsustainable ecosystem type (Vitousek et al. 1997). LUCC not only changes the vegetation cover type and distribution pattern on land surface, but also has an important impact on terrestrial carbon cycle. However, some studies have suggested that the implementation of land use and management measures can make 60–70% of the released carbon refixed and absorbed by terrestrial vegetation (Lal 2002). LUCC has reduced the carbon sequestration potential of ecosystems by 5% in the last 20 years of the twentieth century (DeFries et al. 1999). Therefore, the unreasonable LUCC profoundly affects the material and energy cycle of the ecosystem (Imhoff et al. 2004).

In order to alleviate grassland degradation, restore the ecological environment barrier of grassland, actively play the role of grassland carbon sequestration, and reduce the damage of sandstorms to the natural environment and the serious impact on human survival and life. The Chinese government has implemented a nationwide large-scale ecological restoration project, such as the project of returning farmland to forests and grasslands, which was implemented in 1999, and the implementation of the project of returning grazing to grassland in 2003. The implementation of the two major ecological restoration projects led to changes in land use types, grassland cover areas and vegetation community structures, and grassland production capacity and ecological service functions. The implementation of ecological measures promotes the mitigation of grassland grazing pressure and is conducive to the increase of soil organic carbon in grassland (Wang et al. 2011). Especially in key implementation areas, such as key counties and counties in Inner Mongolia, Shaanxi, and Ninxia, the effects of land desertification and grassland degradation are significant, and sandstorms are controlled to some extent (Wang et al. 2012).

The driving mechanism of grassland degradation is complex; however, overloaded grazing is considered to be the leading driver of grassland degradation (Peng 1993). With the application of remote sensing technology, grassland degradation remote sensing monitoring has higher effectiveness and reliability than traditional field observation methods (Alfredo et al. 2002; Lu et al. 2007) and has high spatial resolution and temporal resolution, which makes grassland degradation dynamic monitoring and degradation classification more efficient and convenient. However, previous researches on the status and driving forces of grassland degradation have focused on grasslands in northern China, typical grasslands in Inner Mongolia, and alpine grasslands on the Qinghai–Tibet Plateau (Li 1997; Liu et al. 2004). There are few studies on grassland degradation and degradation dynamics across the country and for long periods of time. Insufficient quantitative evaluation of the driving mechanism, due to differences in degradation indicators and research methods, and differences in research areas and time periods, leads to differences in grassland degradation and land desertification driving (Wang et al. 2006; Zheng et al. 2006). Therefore, the quantitative evaluation method of grassland degradation driving mechanism is constructed, and the dominant factors of grassland degradation in different regions of China are identified. This is of great significance for the evaluation of grassland degradation and the effective implementation of ecological restoration measures in China.

In summary, in the context of the implementation of global climate change and ecological restoration projects, we will study the grassland productivity and grassland carbon sinks dynamics in China for the past 30 years, analyze the impact of climate change and human activities on grassland NPP, simulate the accounting of grassland carbon sinks in China, explore the dynamics of grassland degradation, quantitatively assess the driving contribution of climate change and human activities to grassland degradation, and clarify the dominant drivers of grassland degradation in different regions. This is of great theoretical significance for further promoting the rational and effective implementation of ecological restoration projects, actively exerting the carbon sequestration potential of grassland in China, and promoting the rational use of grassland resources.

1.2 Related Definitions and Model Formulation

1.2.1 Basic Process of Carbon Cycle

As one of the most widely distributed types of vegetation in the world, grassland accounts for 20% of the global land area; the huge distribution area makes grassland play an important role in the terrestrial carbon cycle. Carbon storage in grassland ecosystems and soil accounts for 34–37% of global terrestrial carbon storage (Matthews et al. 2000). Therefore, the assessment of grassland ecosystem carbon cycle and carbon sinks function is a key link in understanding terrestrial ecosystems and even the global carbon cycle.

Plants, litter, and soil humus constitute the three major carbon pools of grassland ecosystems. While the grassland ecosystem carbon cycle is carried out in the atmosphere, grassland vegetation, and soil (Yu et al. 2003), the carbon stocks and carbon fluxes in each carbon pool are estimated, and the study of its changing trend is the core of the whole grassland carbon cycle research (Qi et al. 2003).

For grassland ecosystems, green plants convert atmospheric CO_2 into organic matter through photosynthesis and store it in plants. This process is an important part of carbon input from the atmosphere to the grassland ecosystem. It is also the material and energy basis of the carbon cycle of grassland ecosystems, which include both above-ground and underground parts (Tao et al. 2001). Due to the wide distribution and diversity of grassland ecosystems, there is significant spatial heterogeneity in grassland vegetation composition, soil, and climatic conditions. Therefore, there are also large differences in the factors affecting the grassland carbon cycle.

Among them, temperature, precipitation, atmospheric CO_2 concentration, growth season length, and soil texture all play an important role in driving grassland carbon cycle. These factors indirectly affect grassland ecosystem carbon by affecting plant community composition, plant physiological characteristics, and hydrothermal conditions (Zhang et al. 2006). Human factors such as grazing and reclamation have significant interference effects on grassland coverage, vegetation growth conditions, and soil nutrient turnover rate, which in turn affects the formation and distribution of grassland net primary productivity (Niu 2001).

Part of the grassland vegetation primary production of carbon is eaten by herbivores, eaten by the animal, part of which is used to maintain its own metabolism and complete secondary production, and the other part is returned to the soil in the form of feces. The aerial part of the plant that is not eaten by animals inputs carbon into the soil by forming litter, and the underground part inputs carbon into the soil by forming plant roots, which is the main source of soil organic carbon.

Carbon emissions from grassland ecosystems include autotrophic respiration of plants, soil roots, soil microbes, and soil animals. Respiratory grassland soil respiration is an important pathway for CO_2 release from grassland ecosystems. Carbon emissions from grassland ecosystems include autotrophic respiration of plants, heterotrophic respiration of soil roots, soil microbes, and soil animals. Respiratory grassland soil respiration is an important pathway for CO_2 release from grassland ecosystems. Compared with forest and farmland ecosystems, the carbon storage of above-ground biomass of grassland ecosystem is not obvious. Carbon is mainly distributed in soil. The global grassland ecosystem has carbon storage of about 308 Pg, of which about 92% is stored in soil and less than 10% in above-ground biomass (Yu et al. 2003).

1.2.2 Observation and Simulation of NPP and Carbon Sinks in Grassland Ecosystem

(1) Grassland NPP simulation

With the development of global change research and remote sensing computer technology, the modeling of terrestrial carbon fluxes and reserves has been developed rapidly at regional and global scales, and NPP is the main measure of terrestrial carbon flux among them. The existing NPP estimation models are summarized as four types: climate productivity model, light utilization model, physiological and ecological process model, and ecological remote sensing model (Ruimy et al. 1994).

(a) Climate productivity model: Such models only consider the impact of climatic conditions on productivity in the estimation of NPP. There are many climate productivity models, including Miami model (Lieth and Box 1972), Thornthwaite Memorial model, Chikugo model, and Zhou Guangsheng model (Uchijima and Seino 1985).

(b) Physio-ecological process model: A physiological–ecological model based on plant growth and development and individual-level dynamics and a simulation model based on the internal function process of ecosystem is to simulate the process of ecosystem structure and functional change on homogeneous patches. The simulated spatial scale is small, neglecting the influence of spatial heterogeneity. This kind of model mainly has compartment models such as CARAIB (Running and Hunt 1993; Warnant et al. 1994), BIOME-BGC (Running and Hunt 1993), CENTURY (Parton et al. 1993); big-leaf model TEM (McGuire et al. 1997), SILVAN (Kaduk and Heimann 1996), and KGBM (Kergoat 1998). The advantages of these models are that they have clear mechanism and can be coupled with atmospheric circulation model, but the disadvantages are that the process model is more complex and needs too many parameters, so it is difficult to be popularized and applied in large-scale space.

(c) Ecological remote sensing coupling model: First, combining the physiological and ecological process model with remote sensing technology, the spatial distribution and dynamic change of NPP on regional and global scales were studied. For example, the NASA earth observation system uses the improved PEM model to estimate NPP for MODIS data. Second, Boreal Ecosystem Productivity Simulator model was developed by combining LAI with remote sensing data, such as the BEPS model based on FOREST-BGC and LAI, and the InTEC model was developed on this basis (Chen et al. 2000).

(d) Light utilization model.

Estimation of vegetation NPP using the light energy utilization rate model is based on the resource balance theory (Field et al. 1995). Monteith proposed to estimate NPP by using APAR of absorbable photosynthetic effective radiation of vegetation and the utilization ratio of light energy, which is known as Monteith's equation.

$$NPP = APAR \times \varepsilon$$

In the formula, ε, the light energy utilization rate of plants is affected by the conditions of water, temperature, and nutrients.

In recent years, the model of light energy utilization has been widely applied in global NPP simulation. The main models in this area are CASA (Potter et al. 1993), GLO-PEM (Prince and Goward 1995), SDBM (Knorr and Heimann 1995), and so on.

(2) **Carbon sinks observation and simulation**

Net ecosystem productivity (NEP) is the result of a balance between the two basic physiological processes of photosynthesis and respiration in the ecosystem (Valentini et al. 2000). At present, the estimation of NEP mainly includes two ways: On the one hand, it can be obtained by model simulation; on the other hand, it can be obtained by observation method. Biome-BGC model, Century model, and TEM model are widely used in the simulation of carbon flux at small homogeneous plot level, and the dynamic changes of ecosystem NEP can be obtained on both the day scale and the year scale. Eco-remote sensing coupling models such as BEPS model, LPJ model, and InTEC model are applied to the simulation and dynamic study of ecosystem carbon flux at regional and global scales. InTEC model can also predict the future changes of NPP and NEP in ecosystem carbon flux.

The methods of observing carbon flux NEP in terrestrial ecosystem mainly include the following four methods: chemical method, box method, micrometeorology method, and soil concentration profile method. Micrometeorological method and box-type method are the two most widely used methods at present (Burkart et al. 2007). Box method has been widely used in grassland, farmland, and forest ecosystems in China (Zhou et al. 2004). Micrometeorology is a method of calculating carbon and water fluxes by measuring gas concentration changes and near-surface turbulence. It mainly includes energy balance method, mass balance method, and aerodynamic method and eddy covariance method. The eddy covariance method (EC) is a method to calculate the material and energy exchange between atmosphere and vegetation by measuring the fluctuation of gas and wind velocity at a certain height. Compared with conventional methods, the vorticity correlation technique has higher measurement accuracy, and the sampling area of this method ranges from hundreds of meters to thousands of meters longitudinally (Schmid 1994). On the time scale, the method can measure carbon and water fluxes of ecosystems on the scale of half an hour to several years (Baldocchi et al. 2001). With the establishment of global flux websites, vorticity correlation method has gradually become the standard method for direct measurement of carbon flux at the ecosystem level, while box method is still widely used in small scale.

1.3 Literature Review

1.3.1 Overview of Grassland Degradation

The concept of grassland degradation has many definitions. In recent decades, domestic scholars have carried out a lot of research on grassland degradation. Li Bo pointed out that grassland degradation is a state in which grassland ecosystem deviates from top-level succession under artificial activities such as grazing, reclamation, and logging (Li 1990). Huang Wenxiu believes that grassland degradation is a process of livestock production decline due to the decrease in livestock carrying capacity (Huang 1991). Some scholars believe that grassland degradation is the main form of land desertification. It is the degradation of grassland vegetation and soil quality caused by disturbance of human activities and unfavorable changes of natural environment. It is also the reduction of productivity, livestock potential and ecological services, environmental degradation and biodiversity, and weakening of ecological restoration (Li 1997).

Grassland degradation is a universal problem in the world, and grassland degradation has become a serious ecological environment problem facing the international community. It is estimated that about 20% of grassland biomass has declined worldwide, with the largest area of grassland degradation in Asia being 3.7 million km^2, accounting for 22% of the total grassland area. The grassland in China has become one of the most severely degraded areas in the world because of its complex topography, sensitive to climate change, and fragile ecological environment. China's grassland degradation began in the 1960s. In the mid-1970s, the degraded grassland area accounted for 15% of the total grassland area. In the 1980s, with the rapid increase of population, the increase of grain and livestock products and the expansion of cities, grassland degradation caused by overloading grazing and grassland reclamation became more and more serious. In the middle 1980s, the area of grassland degradation accounted for 30% of the total grassland area, and in the middle 1990s, it reached more than 50%. The ecological environment of grassland showed an increasingly deteriorating trend (Fan et al. 2007). It has been reported that 90% of the grasslands in China have been degraded to varying degrees. The area of moderate and severe degraded grasslands is 1.3 million km^2, and the rate of degradation is 6700 km^2 per year (Ren et al. 2007). The above-ground biomass of grassland vegetation has decreased from 2.2 to 3.0 t/ha in the 1950s to 0.7–0.9 t/ha in the 1990s (Akiyama and Kawamura 2007).

There are great differences in grassland degradation in different regions. The degraded grassland area of Inner Mongolia is reported to be about 250,000 km^2, accounting for 39.2% of the available grassland area, and the degraded area is increasing at 830 km^2/year rate. In the western part of Northeast China, the area of ungraded grassland is less than 26%, i.e., most grasslands are in the state of degradation, of which the area of mild, moderate, and severe degraded grassland is 4.635 million hectares, 4.81 million hectares, and 9.030 million hectares, respectively (Zhang et al. 2011). Yunnan–Guizhou Plateau and Qinghai–Tibet Plateau are facing severe water erosion and wind erosion. The proportion of edible grass is only 36% (Ren et al. 2007). Degraded grassland in Xinjiang has become a serious disaster area of ecological environment in Xinjiang. 80% of grassland in the whole region has been degraded to varying degrees, and the yield of grassland has decreased by 30–60% (Jia 2007).

1.3.2 Grassland Degradation Monitoring Method

The research on grassland resource management and grassland degradation started late in China. In the early stage of grassland degradation research, ground survey was the main method. With the development of remote sensing technology and the improvement of grassland resource management technology, remote sensing methods were widely used in grassland degradation monitoring and investigation. The traditional field survey of grassland is time-consuming, laborious, inefficient, and the wide distribution of grassland area makes the drawing of grassland degradation grade map and the precision of survey results lower (Asrar et al. 1986). In contrast, the remote sensing method for grassland degradation has high temporal and spatial resolution and fast data updating. Therefore, the results of remote sensing monitoring and grading of grassland degradation have higher reliability, which makes it widely used in grassland degradation monitoring (Alfredo et al. 2002).

According to the connotation and characteristics of grassland degradation, the remote sensing diagnostic index and the classification standard of grassland degradation grade were established. Grassland degradation involves changes in vegetation and soil physiological characteristics of grassland ecosystem. According to the Technical Regulations for Monitoring Grassland Resources and Ecology designated by the Ministry of Agriculture in 2006, ten indicators were selected to classify the degraded grassland into four grades. The ten indices were as follows: the decline rate of plant community biomass, the decay rate of dominant plant population, the decrease rate of dominant grass population yield, the decrease rate of edible herbage yield, the increase proportion of poisonous weeds in grassland community, the decrease percentage of grassland coverage, the decrease rate of plant height, degradation Indicators Plant Growth Rate, Soil Bare Area Status, Light Soil Erosion Degree, Medium and Heavy Soil Volume Weight, Hardness Increase Rate and Restorable Years (Center 2006).

The degree of grassland degradation can also be judged by indicator plants. Liu Zhongling and others have made extensive investigations and observed the succession of grassland communities in Inner Mongolia grassland area for many years. Indicator plant species at different succession stages of grassland degradation have been obtained (Liu et al. 2002).

1.3.3 Driving Factors of Grassland Degradation

There are many factors affecting the degradation of grassland ecosystem—natural factors such as long-term drought, wind erosion, water erosion, air temperature, and precipitation(164), and human factors such as overgrazing, mowing, grassland reclamation, cutting, mineral mining—and grassland input is insufficient. Natural factors have long-term, irreversible, and gradual effects on grassland ecosystem, while human activities have occasional and destructive effects, and it is dominant and controlling to grassland degradation. It is generally believed that climate change and other natural factors are the basis of regional ecological environment changes, and human disturbance activities cause system dysfunction and aggravate grassland degradation, so human activities are induced factors. In general, human factors are the main causes of grassland degradation (Meinzen-Dick and Di Gregorio 2004). The driving forces of grassland degradation in China are mainly climate change (including temperature and precipitation) and overgrazing (Akiyama and Kawamura 2007; Li et al. 2008; Liu and Wang 2007; Ren et al. 2007).

Grazing is one of the main ways of grassland resource utilization and is also the main factor of grassland degradation. About 35% of the degraded grasslands in the world are caused by overgrazing, which accounts for more than 20% in China (Yu et al. 2003). As far as the degree of impact is concerned, it is much higher than grassland reclamation.

Grassland reclamation, for a long time, has been regarded as a suitable agricultural wasteland with flat land, fertile soil and good grassland vegetation growth, and is regarded as the object of reclamation. Since 1949, under the guidance of the "grain-oriented" policy, the whole country has set off a number of reclamation tides in grassland areas. During the 50 years from 1949 to 1999, four major reclamations in China have led to the reclamation of 193,000 km^2 of grassland into farmland. At present, 18.2% of the increase in farmland area in China is due to the reclamation of grassland (Han et al. 2008). Large-scale reclamation of grassland will lead to the rapid expansion of sandy desertification (Zhu 1997). At the same time, grassland reclamation places, mostly located in the upper reaches of the river system or the birthplace, reclamation of vegetation has been destroyed, surface bare, soil erosion intensified. When rainfall is too small, runoff will decrease and even rivers and lakes will be cut off. With the increase of the number of gale days, more sand and dust storms are formed.

Large-scale logging and random digging have seriously affected grassland vege-tation, accelerating grassland degradation and desertification, and become one of the causes of grassland degradation(Chen 1990). It is reported that in the desert steppe of Ordos, Inner Mongolia, pits are everywhere for digging medicinal herbs such as licorice and ephedra. Each kilogram of licorice is dug to destroy 0.53–0.67 hectares of grassland. The area of grassland destroyed by digging medicinal herbs is 26,700 hectares annually (Li 1997). In addition, woodcutting fuel-wood also accelerated grassland degradation. For example, in the 1970s and 1980s, farmers and herdsmen in Ordos, Inner Mongolia, cut down a large number of shrubs in order to solve the problem of firewood, resulting in the destruction of about 40 mu of grassland per household per year for firewood only (Bao 2003). In addition, inappropriate mining practices exacerbated grassland degradation to a certain extent (Xu and Zhao 2000). Overgrazing and grassland reclamation are the main human driving factors of grass-land degradation, and they also have an important impact on the change of grassland carbon storage (Rodriguez-Murillo 2001; Su et al. 2006).

Rodent and insect pests on grasslands are also important causes of grassland degradation. It is reported that the habitat density and the hole density of voles in typical grassland areas of Inner Mongolia are 384 voles/ha and 6920 holes/ha, respec-tively. Their feeding and digging on vegetation will reduce the primary production of grassland vegetation and lead to soil degradation. The loss of forage is as high as 44% in severe disaster years and 15–20% in normal years (Zhang and Yang 2011).

Climate change is also one of the factors leading to grassland degradation. In recent years, with the global warming, the impact of climate change on grassland ecosystem has become increasingly serious, which has attracted the attention of the international community (Nautiyal et al. 2004). As temperature increases, above-ground biomass, edible forage biomass, and plant species diversity will be reduced (Klein et al. 2004, 2007). The aridity of climate and the in-homogeneity of spatial and seasonal distribution of precipitation have restrained the growth of grassland vegetation to some extent, and even resulted in the succession of grassland degradation. The study shows that the temperature has increased significantly in the past 1982–1999 years, and the precipitation has changed little (Piao et al. 2004).

1.3.4 Ecological Restoration Project

Most of the grasslands in China are located in arid and semi-arid climatic regions. The ecological environment is fragile and sensitive to climate change. At the same time, overexploitation of resources, exploitation and human disturbance, have further aggravated the degradation of ecological environment. China has become one of the countries with the most serious degradation of grasslands in the world. In the past 30 years of reform and opening up, China's economy has developed at an unprecedented speed, accompanied by the accelerated deterioration of the ecological environment. At present, soil erosion, land desertification, soil salinization, sandstorms, and the reduction of biodiversity are dominant ecological and environmental problems in China. In addition, the severe drought in 1997 and the flood and waterlogging in 1998 have made people strongly aware that speeding up the construction of forest and grass vegetation and improving the ecological environment have become an urgent task facing the people of the whole country. Therefore, since the end of twentieth Century, a series of large-scale ecological restoration projects have been implemented by our government. For example, the Natural Forest Protection Project (which began in 1998 and was piloted in 12 provinces with a planned investment of 96.2 billion yuan in 2000–2010), the Project of Returning Farmland to Forests and Grassland (with a total investment of 220 billion yuan in 1999–2010), the Project of Returning Grassland to Grassland (with a total investment of 13.57 billion yuan in 2003–2010), "Beijing-Tianjin Sandstorm Source Control Project" (2000–2010, the central accumulative investment of 4.4 billion yuan), and "Southwest Karst Region Grassland Control Pilot Project" (2006–2010, the central accumulative investment of 137 million yuan), and so on. These measures have a significant impact on the ecology and social economy of grassland distribution areas (Wu and Cai 2009) and have a great impact on land use and cover and terrestrial carbon cycle in China. At the same time, they have also played a positive role in promoting the socioeconomic development and environmental improvement of the surrounding areas and have achieved better ecological, social, and economic benefits. Among them, the project of returning farmland to forest and grassland and the project of returning pasture to grassland have the most obvious effect in terms of scope, intensity, investment, and restoration.

1.4   Problems in Carbon Sinks Accounting and Grassland Degradation Research

1.4.1   Studies on the Driving Mechanism of Grassland NPP in China Are Few

Previous studies on grassland NPP were mainly based on small-scale regional or sample plot level, but there was little overall study on grassland ecosystem productivity at long-term scale in China. At the same time, because of the difference of different simulation methods and parameters setting, the comparability of NPP between regions is often poor. In addition, the impact of driving factors on NPP was mostly discussed from a single climate or land-use/cover change factor, while the coupling of LUCC and grassland NPP was seldom studied. However, grasslands are widely distributed and varied in China, and the control factors such as climate, soil, and human activities are quite different in different regions and grasslands. The results of regional studies only reflect the differences of regional conditions and cannot comprehensively and objectively reveal the size and spatial distribution characteristics of grassland productivity on the national scale. Therefore, the estimation of grassland NPP at the national scale is helpful to better evaluate the carbon sequestration capacity of grassland ecosystem in China, and the coupling study of driving mechanism is also helpful to rational utilization and effective management of grassland resources.

1.4.2   Research on the Grassland Carbon Sinks Evaluation in China's National Scale Area Few

Grassland NEP is an important measure index of carbon sinks function. Previous studies on grassland NEP in China were mostly confined to small scales, such as sample plot level or grassland type, but the study on grassland NEP accounting and spatial distribution pattern at national scale was insufficient. In addition, in the simulation and observation methods of carbon flux NEP, ecological process model is used to simulate grassland carbon flux at homogeneous sampling points. Ground observation data are scarce, and long-term positioning observation stations and flux stations are concentrated on the typical grassland of Inner Mongolia and alpine grassland of Qinghai–Tibet Plateau. The data can be used to verify the accuracy of NEP model simulation and cannot fully reflect the carbon sinks function of grassland in China. Therefore, the high spatial and temporal resolution of carbon flux data can be obtained by using the ecological remote sensing coupling model. Comprehensive assessment of carbon sinks function and its driving mechanism of grassland in China is of great significance to clarify carbon budget and carbon cycle of grassland in China.

1.4.3 Lack of Remote Sensing Monitoring and Driving Mechanism Quantitative Assessment of Grassland Degradation in China

Traditional grassland degradation assessment is mostly based on field investigation, with low efficiency and low accuracy. However, current monitoring tasks for grassland degradation monitoring by remote sensing: in the research scale, mostly concentrated in the regional scope, less research on the national scale; research content, limited to the classification of grassland degradation, grassland degradation spatial distribution pattern of less research; in the selection of degradation indicators, the indicators selected by different researchers are not uniform, resulting in poor comparability of degradation status. In addition, the research on driving mechanism of grassland degradation is mostly qualitative analysis or statistical analysis based on socioeconomic indicators. The quantitative evaluation of driving force using remote sensing technology is less. Although climate factors and human activities are the two main driving factors of grassland degradation, the quantitative study on their contribution to grassland degradation is still blank, and the grassland degradation dominant factors in different regions are not clear. Therefore, the quantitative and spatial study on the driving mechanism of grassland degradation is of great significance to the sustainable utilization of grassland resources and the rational implementation of the restoration measures of degraded grassland. In addition, determining the size and specialization of contribution which climate and human factors are making to grassland degradation can reflect the implementation performance of ecological restoration measures to a certain extent.

References

Akiyama T, Kawamura K (2007) Grassland degradation in China: methods of monitoring, management and restoration. Grassland Sci 53(1):1–17

Alfredo CD, Emilio CW, Ana C (2002) Satellite remote sensing analysis to monitor desertification processes in the crop-rangeland boundary of Argentina. J Arid Environ 52(1):121–133

Asner GP, Martin RE (2004) Biogeochemistry of desertification and woody encroachment in grazing systems. Ecosyst land use change 99–116

Asrar G, Weiser RL, Johnson DE, Kanemasu ET, Killeen JM (1986) Distinguishing among tallgrass prairie cover types from measurements of multispectral reflectance. Remote Sens Environ 19(2):159–169

Baldocchi D, Falge E, Gu L, Olson R, Hollinger D, Running S (2001) FLUXNET: a new tool to study the temporal and spatial variability of ecosystem-scale carbon dioxide, water vapor, and energy flux densities. Bull Am Meteor Soc 82(11):2415–2434

Bao YS (2003) The history and future of grassland animal husbandry in Inner Mongolia. Inner Mongolia Education Press, Hohhot

Burkart S, Manderscheid R, Weigel H (2007) Design and performance of a portable gas exchange chamber system for CO_2 and H_2O flux measurements in crop canopies. Environ Exp Bot 61(1):25–34

Chen ZZ (1990) Degradation and regulation of natural grassland ecosystem in China. Study on land degradation control in China. China Science and Technology Press, Beijing, pp 86–88

Chen WJ, Chen J, Cihlar J (2000) An integrated terrestrial ecosystem carbon-budget model based on changes in disturbance, climate, and atmospheric chemistry. Ecol Model 135(1):55–79

Conant RT, Paustian K, Elliott ET (2001) Grassland management and conversion into grassland: effects on soil carbon. Ecol Appl 11(2):343–355

DeFries RS, Field CB, Fung I, Collatz GJ, Bounoua L (1999) Combining satellite data and bio-geochemical models to estimate global effects of human-induced land cover change on carbon emissions and primary productivity. Glob Biogeochem Cycles 13(3):803–815

Fan JW, Zhong HP, Chen LB, Zhang WY (2007) Some scientific problems of grassland degradation in arid and semi-arid regions in northern China. Chin J Grassland 29(5):95–101

Field CB, Randerson JT, Malmström CM (1995) Global net primary production: combining ecology and remote sensing. Remote Sens Environ 51(1):74–88

Han JG, Zhang YJ, Wang CJ, Bai WM, Wang YR, Han GD (2008) Rangeland degradation and restoration management in China. Rangeland J 30(2):233–239

Houghton JT, Ding Y, Griggs DJ, Noguer M, Linden PJ, Dai X (2001) Climate change 2001: the scientific basis. Cambridge University Press, Cambridge

Huang WX (1991) Development of animal husbandry resources and base construction in Southwest China: Beijing. Science Press

Imhoff ML, Bounoua L, DeFries R, Lawrence WT, Stutzer D, Tucker CJ (2004) The consequences of urban land transformation on net primary productivity in the United States. Remote Sens Environ 89(4):434–443

Jia HT (2007) Ecological effects of enclosure on degraded grassland in Xinjiang. Xinjiang Agricultural University

Kaduk J, Heimann M (1996) A prognostic phenology scheme for global terrestrial carbon cycle models. Climate Res 6(1):1–19

Kergoat L (1998) A model for hydrological equilibrium of leaf area index on a global scale. J Hydrol 212:268–286

Klein JA, Harte J, Zhao XQ (2004) Experimental warming causes large and rapid species loss, dampened by simulated grazing, on the Tibetan Plateau. Ecol Lett 7(12):1170–1179

Klein JA, Harte J, Zhao X (2007) Experimental warming, not grazing, decreases rangeland quality on the Tibetan Plateau. Ecol Appl 17(2):541–557

Knorr W, Heimann M (1995) Impact of drought stress and other factors on seasonal land biosphere CO_2 exchange studied through an atmospheric tracer transport model. Tellus B 47(4):471–489

Lal R (2002) Soil carbon sequestration in China through agricultural intensification, and restoration of degraded and desertified ecosystems. Land Degrad Dev 13(6):469–478

Levy PE, Cannell M, Friend AD (2004) Modelling the impact of future changes in climate, CO_2 concentration and land use on natural ecosystems and the terrestrial carbon sink. Glob Environ Change 14(1):21–30

Li B (1990) Study on natural resources and environment of Ordos Plateau in Inner Mongolia. Science Press, Beijing

Li B (1997) The degradation of grassland in North China and its countermeasure. Agr Sci Sin 30:1–10

Li XL, Yuan QH, Wan LQ, He F (2008) Perspectives on livestock production systems in China. Rangeland J 30(2):211–220

Lieth H, Box E (1972) Evapotranspiration and primary productivity. Publ Climatol 25(2):37–46

Liu SL, Wang T (2007) Aeolian desertification from the mid-1970s to 2005 in Otindag Sandy Land, Northern China. Environ Geol 51(6):1057–1064

Liu ZL, Wang W, Hao DY, Liang CZ (2002) Prebes on the degeneration and recovery succession mechanisms of Inner Mongolia Steppe. J Arid Resour Environ 16(1):84–91

Liu Y, Zha Y, Gao J, Ni S (2004) Assessment of grassland degradation near Lake Qinghai, West China, using Landsat TM and in situ reflectance spectra data. Int J Remote Sens 25(20):4177–4189

Lu D, Batistella M, Mausel P, Moran E (2007) Mapping and monitoring land degradation risks in the Western Brazilian Amazon using multitemporal Landsat TM/ETM + images. Land Degrad Dev 18(1):41–54

Matthews E, Payne R, Rohweder M, Murray S (2000) Pilot analysis of global ecosystems: Forest ecosystems. World Resources Institute, Washington DC

McGuire AD, Melillo JM, Kicklighter DW, Pan Y, Xiao X, Helfrich J (1997) Equilibrium responses of global net primary production and carbon storage to doubled atmospheric carbon dioxide: Sensitivity to changes in vegetation nitrogen concentration. Global Biogeochem Cycles 11(2):173–189

Meinzen-Dick RS, Di Gregorio M (2004) Collective action and property rights for sustainable development. International Food Policy Research Institute, Washington

Milchunas DG, Lauenroth WK (1993) Quantitative effects of grazing on vegetation and soils over a global range of environments. Ecol Monogr 63(4):327–366

Millington JDA, Perry GLW, Romero-Calcerrada R (2007) Regression techniques for examining land use/cover change: a case study of a Mediterranean landscape. Ecosystems 10(4):562–578

Milton SJ, du Plessis MA, Siegfried WR (1994) A conceptual model of arid rangeland degradation. Bioscience 44(2):70–76

National Animal Husbandry Center (2006) Technical specification for monitoring grassland resources and ecology. NY/T 1233–2006

Nautiyal MC, Nautiyal BP, Prakash V (2004) Effect of grazing and climatic changes on alpine vegetation of Tungnath, Garhwal Himalaya, India. Environmentalist 24(2):125–134

Ni J (2002) Carbon storage in grasslands of China. J Arid Environ 50(2):205–218

Niu JM (2001) Impacts prediction of climatic change on distribution and production of grassland in inner Mongolia. Acta Agrestla Sin 9(4):277–282

Parton WJ, Scurlock JMO, Ojima DS, Gilmanov TG, Scholes RJ, Schimel DS (1993) Observations and modeling of biomass and soil organic matter dynamics for the grassland biome worldwide. Glob Biogeochem Cycles 7(4):785–809

Peng X (1993) Grassland resources and its application of Xinjiang. Xinjiang Science and Technology and Health Press, Ulu

Piao SL, Fang JY, Ji W, Guo QH, Ke JH, Tao S (2004) Variation in a satellite-based vegetation index in relation to climate in China. J Veg Sci 15(2):219–226

Potter CS, Randerson JT, Field CB, Matson PA, Vitousek PM, Mooney HA (1993) Terrestrial ecosystem production: a process model based on global satellite and surface data. Global Biogeochem Cycles 7(4):811–841

Prince SD, Goward SN (1995) Global primary production: a remote sensing approach. J Biogeogr 20(4):815–835

Qi YC, Dong YS, Geng YB, Yang XH, Geng HL (2003) The progress in the carbon cycle researches in grassland ecosystem in China. Prog Geogr 22(4):342–352

Ren H, Shen WJ, Lu HF, Wen XY, Jian SG (2007) Degraded ecosystems in China: status, causes, and restoration efforts. Landscape Ecol Eng 3(1):1–13

Ren JZ, Hu ZZ, Zhao J, Zhang DG, Hou FJ, Lin HL (2008) A grassland classification system and its application in China. Rangeland J 30(2):199–209

Rodriguez-Murillo JC (2001) Organic carbon content under different types of land use and soil in peninsular Spain. Biol Fertil Soils 33(1):53–61

Ruimy A, Saugier B, Dedieu G (1994) Methodology for the estimation of terrestrial net primary production from remotely sensed data. J Geophys Res 99(D3):5263–5283

Running SW, Hunt ER (1993) Generalization of a forest ecosystem process model for other biomes, BIOME-BGC, and an application for global-scale models. In: Scaling physiological processes: leaf to globe, pp 141–158

Schmid HP (1994) Source areas for scalars and scalar fluxes. Bound-Layer Meteorol 67(3):293–318

Scurlock J, Hall DO (1998) The global carbon sink: a grassland perspective. Glob Change Biol 4(2):229–233

Sharrow SH, Ismail S (2004) Carbon and nitrogen storage in agroforests, tree plantations, and pastures in western Oregon, USA. Agrofor Syst 60(2):123–130

Snyman HA, Fouché HJ (1991) Production and water-use efficiency of semi-arid grasslands of South Africa as affected by veld condition and rainfall. Water SA 17(4):263–268

Su YZ, Li YL, Zhao HL (2006) Soil properties and their spatial pattern in a degraded sandy grassland under post-grazing restoration, Inner Mongolia, northern China. Biogeochemistry 79(3):297–314

Tao B, Ge QS, Li KR, Shao XM (2001) Progress in the studies on carbon cycle in terrestrial ecosystem. Geogr Res 20(5):564–575

Turner BL II, Villar SC, Foster D, Geoghegan J, Keys E, Klepeis P (2001) Deforestation in the southern Yucatan peninsular region: an integrative approach. For Ecol Manag 154(3):353–370

Uchijima Z, Seino H (1985) Agroclimatic evaluation of net primary productivity of natural vegetations, 1: Chikugo model for evaluating net primary productivity. J Agric Meteorol 40:343–352

Valentini R, Matteucci G, Dolman AJ, Schulze E, Rebmann C, Moors EJ (2000) Respiration as the main determinant of carbon balance in European forests. Nature 404(6780):861–865

Vitousek PM, Mooney HA, Lubchenco J, Melillo JM (1997) Human domination of Earth's ecosystems. Science 277(5325):494–499

Wang G, Cheng G, Shen Y (2002) Soil organic carbon pool of grasslands on the Tibetan Plateau and its global implication. J Glaciol Geocryol 24(6):693–700

Wang XM, Chen FH, Dong ZB (2006) The relative role of climatic and human factors in desertification in semiarid China. Glob Environ Change 16(1):48–57

Wang X, Zang S, Na X (2011) Analyzing dynamic process of land use change in Ha-Da-Qi industrial corridor of China. Procedia Environ Sci 11(Part B):1008–1015

Wang T, Sun JG, Han H, Yan CZ (2012) The relative role of climate change and human activities in the desertification process in Yulin region of northwest China. Environ Monit Assess 184(12):7165–7173

Warnant P, François L, Strivay D, Gérard JC (1994) CARAIB: a global model of terrestrial biological productivity. Global Biogeochem Cycles 8(3):255–270

Wu DD, Cai YL (2009) Evaluation of ecological restoration effects in China: A review. Prog Geogr 28(4):622–628

Xu ZX, Zhao ML (2000) Eco-environmental deterioration and strategies for preventing it in Inner Mongolia. Grassland China (5):59–63

Yu GR, Li HT, Wang SQ (2003) Global change, carbon cycle and storage in terrestrial ecosystem. Meteorological Press, Beijing

Zhang WH, Yang W (2011) The feature analysis for grassland degradation and the restoration of natural vegetation in degraded grassland. Northern Environ (8):40–44

Zhang W, Zhang H, Ze B (2006) Progress studies on the carbon cycle of Alpine meadow in China. J Mt Sci 24(B10):266–274

Zhang CX, Wang XM, Li JC, Hua T (2011) Roles of climate changes and human interventions in land degradation: a case study by net primary productivity analysis in China's Shiyanghe Basin. Environ Earth Sci 64(8):2183–2193

Zheng YR, Xie ZX, Robert C, Jiang LH, Shimizu H (2006) Did climate drive ecosystem change and induce desertification in Otindag sandy land, China over the past 40 years. J Arid Environ 64(3):523–541

Zhou CY, Zhang DQ, Wang SY, Zhou GY, Liu SZ, Tang XL (2004) Diurnal variations of fluxes of the greenhouse gases from a coniferous and broad-leaved mixed forest soil in Dinghushan. Acta Ecol Sin 24(8):1738–1741

Zhu ZD (1997) Global change and desertification. Earth Sci Front 4(1):213–219

Chapter 2
Grassland Coverage Inter-Annual Variations and Its Coupling Relation with Hydrothermal Factors in China

Abstract Global inventory modeling and mapping studies (GIMMS) and normalized difference vegetation index (NDVI), from 1982 to 2010, were used to simulate the grass coverage and analyze its spatial pattern and changes. The response of grass coverage to climatic variations at annual and monthly time scales was analyzed (during the 29 years, the nationwide annual temperature increased with a mean rate of 0.04 °C/year and precipitation decreased with a mean rate of −0.39 mm/year; however, in northwest China, precipitation increased). Grass coverage distribution had increased from northwest to southeast across China. During 1982–2010, the mean nationwide grass coverage was 34%, but exhibited apparent spatial heterogeneity being the highest (61.4%) in slope grasslands and the lowest (17.1%) in desert grasslands. There was a slight increase in the grass coverage over the study period with a rate of 0.17% per year. Regionally, the largest increase of grass coverage was observed in northwest China and Tibetan Plateau. Increase in slope grasslands coverage was as high as 0.27% per year, while in the plain grasslands and meadows, the grass coverage increase was the lowest (being 0.11% per year and 0.1% per year, respectively). Across China, the grass coverage with extremely significant increase ($P < 0.01$) and significant increase ($P < 0.05$) accounted for 46.03% and 11% of the total grassland area, respectively, while those with extremely significant and significant decrease accounted for only 4.1% and 3.24%, respectively. At the annual time scale, there are no significant correlations between grass coverage and annual mean temperature and precipitation for the total grassland area. However, the grass coverage was somewhat affected by temperature in alpine and sub-alpine grassland, alpine and sub-alpine meadow, slope grassland and meadow, while grass coverage in desert grassland and plain grassland was more affected by precipitation. At the monthly time scale, there are significant correlations between grass coverage with both temperature and precipitation, indicating that the grass coverage is more affected by seasonal fluctuations of hydrothermal conditions. Additionally, there is one-month time-lag effect between grass coverage and climate factors for each grassland types, and the correlations are the highest between the current months' grass coverage and the former one month's temperature and precipitation.

Keywords Grass coverage · Spatiotemporal dynamic · Climate factors · Correlation · Time-lag effect

© Springer Nature Singapore Pte Ltd. 2020   17
W. Zhou et al., *Remote Sensing Monitoring and Evaluation of Degraded Grassland in China*, Springer Geography,
https://doi.org/10.1007/978-981-32-9382-3_2

2.1   Introduction

Response of terrestrial ecosystems to global climate change is one of the most complex research topics in the global change studies (Walker and Steffen 1997). Vegetation, as the main component of terrestrial ecosystem, is sensitive to climatic change. Climate change has caused the change of vegetation growth environment, and then impacted vegetation dynamic, composition, and functions (Keeling et al. 1996; Parmesan and Yohe 2003; Roerink et al. 2003; Weltzin et al. 2003; Xin et al. 2008). Vegetation coverage is an important ecological parameter, which reflects the degree of lush vegetation and the photosynthetic area. Its variations are the direct results of regional environmental change (Xin et al. 2008), and it executes an important indication function of regional environment change (Gan et al. 2011). Researches on the response of plant growth to climate change have been widely conducted based on normalized difference vegetation index (NDVI) (Hall et al. 1992; Li et al. 2000). Global warming has led to significant enhancement of vegetation activity in the high latitudes of Northern Hemisphere (Myneni et al. 1997; Zhou et al. 2001). In China, vegetation activity in most regions is also increased, especially in the Tibetan Plateau, northwest China, and north China (Fang et al. 2001; Li et al. 2006; Piao and Guo 2001; Zhang et al. 2006). Although researches on the response of vegetation growth to temperature and precipitation are widely conducted, the conclusions of the correlations between vegetation growth and hydrothermal factors are inconsistent. Some scholars argued that precipitation was an important factor responsible for the vegetation growth and seasonal variations in China (Sun et al. 2013; Tang 2003; Zhao et al. 2001); others found that the effects of temperature to vegetation growth were even larger than precipitation (Cui and Graf 2009; Luo et al. 2009; Sun et al. 1998; Wu and Cai 2009). Response of different vegetation types to hydrothermal factors is different, and the impacts of precipitation and temperature on vegetation coverage are also varied for different spatial and temporal scales and ecosystem types. For example, for the desert grassland in north China, precipitation serves as a restrictive factor (Li et al. 2000).

Grassland is the biggest terrestrial ecosystem type in China and has an important role in the national ecological environment protection plans (Ren et al. 2011). In the recent decades, with the intensification of global climate change and human interference, grassland ecosystem has also significantly changed. At present, studies of the correlation between grass coverage and climate change are mainly focused on selected specific regions and on short-term time scale (Mu et al. 2013). Little research was done on spatial–temporal dynamic of the grass coverage and on its correlation with climate change at national scale and on longer time scale. Grasslands accounting for one-third of China's total land area mainly distributed in the northwest arid and semi-arid climatic zones, and in and around the Tibetan Plateau. These regions are more sensitive to global change (Christensen et al. 2004). Therefore, the research on the grass coverage dynamics and its response to climate change in China will contribute to our understanding of impacts of climate change on terrestrial ecosystem.

In this study, global inventory modeling and mapping studies (GIMMS) and normalized difference vegetation index (NDVI), from 1982 to 2010, were used to calculate grass coverage and their spatial–temporal dynamics. Finally, the correlations between grass coverage and climate factors at annual and monthly time scales were analyzed. This study aims to explore the effects of climate change on vegetation growth and find the climatic restrictive factors for each grassland type. Understanding of these effects provides some theoretical basis and is essential for reliable projections of the grassland ecosystem change in the future climate change.

2.2 Methodology

2.2.1 Normalized Difference Vegetation Indices (NDVI)

The Advanced Very High-Resolution Radiometer (AVHRR) GIMMS NDVI data, with a resolution of 8 × 8 km, covers the periods from January 1982 to December 2010. Data are available at http://westdc.westgis.ac.cn/data/ with 15-day time resolution. During the preparing of the dataset, its creators conducted the radiation correction, geometrical correction, and cloud filtering to improve the data accuracy. At present, this dataset has been widely used in numerous larger-scale land-cover variation studies (Zhang and Yang 2011).

2.2.2 Meteorological Data

Meteorological data from 1982 to 2010, including monthly mean temperature and total precipitation for 720 stations, were obtained from China Meteorological Data Sharing Service System (http://cdc.cma.gov.cn/home.do). Ordinary Kriging interpolation was used to interpolate the meteorological data into grid at 1 × 1 km spatial resolution and with the same coordinate system as the NDVI images. In addition, the error analysis of interpolated meteorological data was conducted in comparison with actual measurements (Table 2.1). Results indicated that there was no significant difference between measured and interpolated data, and the root mean square error (RMSE) of the estimates was less than 1 °C for monthly temperature and 2 mm for monthly precipitation (Table 2.1).

2.2.3 Land-Cover Products

The Global Land Cover 2000 dataset (GLC 2003) with 1 km spatial resolution indicated that China's grassland area accounts for approximately 35% of China's total

Table 2.1 Error analysis for the interpolated meteorological data, using year 2001 as an example

Month	Temperature (°C) ($n = 720$)			Precipitation (mm) ($n = 720$)		
	Mean error	RMSE	R^2	Mean error	RMSE	R^2
Jan	0.006	0.958	0.958	0.289	0.987	0.809
Feb	−0.073	0.985	0.935	0.351	1.425	0.815
Mar	−0.115	0.937	0.859	0.509	0.958	0.931
Apr	−0.136	0.883	0.821	3.149	1.491	0.799
May	−0.124	0.925	0.842	0.438	1.677	0.839
Jun	−0.108	0.960	0.869	1.758	0.975	0.877
Jul	−0.113	0.883	0.894	−10.026	1.709	0.841
Aug	−0.095	0.908	0.886	−7.659	1.506	0.792
Sep	−0.098	0.976	0.893	−4.742	0.997	0.828
Oct	−0.085	0.950	0.897	2.768	1.302	0.817
Nov	0.028	0.875	0.927	0.481	1.146	0.784
Dec	0.008	0.887	0.953	−0.447	1.030	0.803

land area, mainly distributed in the northwest China and the Tibetan Plateau. At present, grassland reclassification accuracy of GLC2000 for China is higher than other land-cover products such as (International Geosphere Biosphere Programme global land-cover datasets (IGBP-DISCover), the University of Maryland (UMD) global land-cover map, and MODIS global land-cover map (Ran et al. 2010). Besides, GLC2000 product has the highest consistency with the China vegetation cover data with 1:100,000 and has six grassland types (Fig. 2.1).

2.2.4 Estimation of Grass Coverage and Its Accuracy Validation

(1) Vegetation coverage estimation

Gutman and Ignatov (1998) developed a semi-empirical relationship between vegetation coverage and NDVI, and proposed a dense vegetation mosaic-pixel model to derive vegetation coverage from NDVI. The model of the calculation of vegetation coverage can be expressed as:

$$\text{NDVI} = \text{NDVI}_v C_i + \text{NDVI}_s (1 - C_i) \tag{1}$$

where NDVI_s is the minimum NDVI corresponding to 0% vegetation cover or bare soil, and NDVI_v is the maximum one with a 100% vegetation cover. The vegetation coverage (C_i) derived by the scaled NDVI can be expressed as:

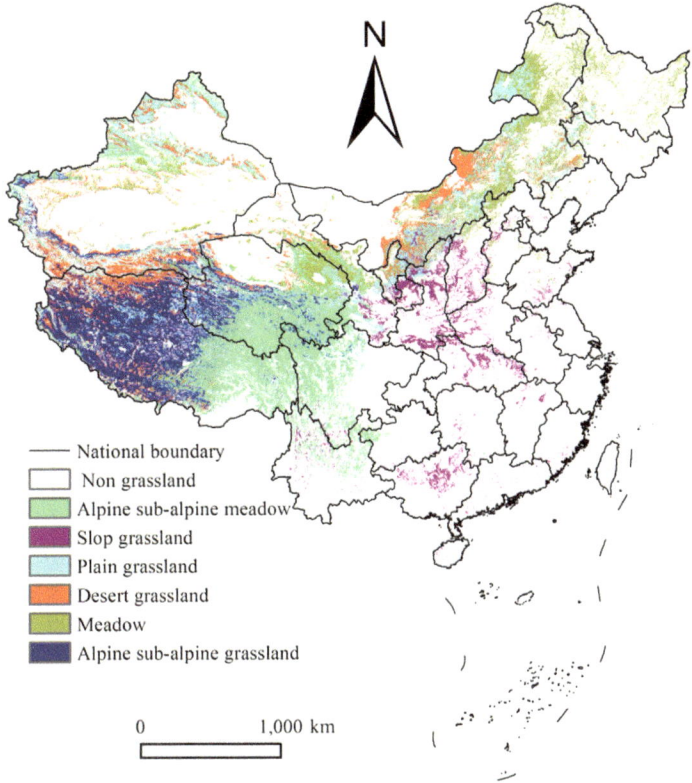

Fig. 2.1 Location of the study area and its grassland-type distribution

$$C_i = \frac{\text{NDVI} - \text{NDVI}_{\text{min}}}{\text{NDVI}_{\text{max}} - \text{NDVI}_{\text{min}}} \qquad (2)$$

where NDVI_{max} and NDVI_{min} stand for the maximum and minimum NDVI of a year, respectively.

(2) Accuracy validation of vegetation coverage

To validate the accuracy of the vegetation coverage, we sampled 90 sites across the study area in July and August of 2009. At each site (10×10 m), five plots (1×1 m) were set, and vegetation coverage was investigated; mean coverage of these five plots was considered as sites vegetation coverage. Correlation between estimated grass coverage and field observation values is shown in Fig. 2.2 ($R^2 = 0.9872$, $P < 0.001$), which indicates that the correlation is significant and the estimated results are reliable.

Fig. 2.2 Correlation analysis of the estimated grass coverage and its field observed values

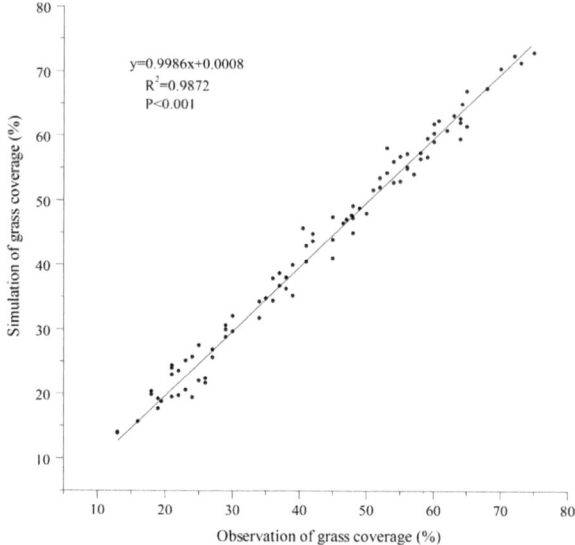

2.2.5   Grassland Vegetation Coverage Trend Analysis

Simple linear regression was used to analyze change trend of vegetation cover from 1982 to 2010. Simple linear regression estimates the change tendency for each grid as follows:

$$\theta_{\text{Slope}} = \frac{n \times \sum_{i=1}^{n} i \times C_i - \left(\sum_{i=1}^{n} i\right)\left(\sum_{i=1}^{n} C_i\right)}{n \times \sum_{i=1}^{n} i^2 - \left(\sum_{i=1}^{n} i\right)^2} \qquad (3)$$

where n is the number of studied years; C_i is the maximum vegetation coverage of year i; θ_{Slope} is the slope of the trend line. To check the significance of the change tendency of coverage, a significance test (F-test) was applied. The F statistic test is shown by Eq. 4:

$$F = U \times \frac{n-2}{Q} \qquad (4)$$

where $U = \sum_{i=1}^{n} (\hat{y}_i - \bar{y})^2$ is the regression sum of squares, $Q = \sum_{i=1}^{n} (y_i - \hat{y}_i)^2$ is residual sum of squares; y_i is the estimated value of vegetation coverage for year i and $\hat{y}_i$ is the regression value; $\bar{y}$ is the average value of annual vegetation coverage from 1982 to 2010; n is the sample size (here, $n = 29$). According to the F-test, the change tendency was classified into six categories: extremely significant decrease (ESD, $\theta_{\text{Slope}} < 0$, $P < 0.01$); significant decrease (SD, $\theta_{\text{Slope}} < 0$, $0.01 < P < 0.05$); no significant change (NS, $P > 0.05$); significant increase (SI, $\theta_{\text{Slope}} > 0$, $0.01 < P < 0.05$); extremely significant increase (ESI, $\theta_{\text{Slope}} > 0$, $P < 0.01$).

2.2.6 Correlation Analysis of Grass Coverage and Climate Factors

To explore the response of grass coverage to climate change, correlation coefficient between vegetation coverage and annual (monthly) mean temperature and total precipitation were calculated by the following equation:

$$R_{xy} = \frac{\sum_{i=1}^{n} [(x_i - \overline{x})(y_i - \overline{y})]}{\sqrt{\sum_{i=1}^{n} (x_i - \overline{x})^2 \sum_{i=1}^{n} (y_i - \overline{y})^2}} \tag{5}$$

where R_{xy} is the correlation coefficient of variable x and y; x_i is the vegetation coverage of the ith year or month; y_i is the temperature or precipitation of the ith year or month; $\overline{x}$ is the average vegetation coverage for all years or months, $\overline{y}$ is the average temperature or precipitation for all years or months; and i is the number of years or months. Applying the non-autocorrelation clause for inter-annual variations of grass coverage, we assume that its major variations occur in the warm season, and therefore, the grass coverage time series should be not autocorrelated at the annual time scale.

2.3 Implementations and Comparisons

2.3.1 Spatial Distribution of Long-Term Mean Grass Coverage

The spatial distribution of long-term mean grass coverage in China is shown in Fig. 2.3a. The grass coverage value is relatively higher in southeast China, while lower in northwest China. For the entire grassland, the mean grass coverage for the 1982–2010 period was 34% with highest in the slope grassland and lowest in desert grassland (Fig. 2.3b).

The slope grasslands are mainly located in the Qinling Mountains, north of Guangxi Zhuang Autonomous Region, and west of Jilin Province, with abundant precipitation and lush vegetation. The mean grass coverage of slope grassland was 61.4% during the 29 years, which ranged from 60 to 80%. The meadow grasslands mainly distributed in the northeast of Inner Mongolia Autonomous Region (IM), west of Heilongjiang and Jilin provinces, as well as the Qilian Mountains region. The mean grass coverage of meadow grasslands was 41.5%, which varies between 40 and 60%. For alpine and sub-alpine meadow grassland type, the mean grass coverage was 40.1%. This grasslands type is located in the south of Tibet Autonomous Region, Qilian Mountains, southern slopes of the Tianshan Mountains, and in the Altai Mountains. Here, the grass coverage decreased from southeast to northwest and ranged from 30 to 60%. Plain grasslands, with the mean vegetation coverage of

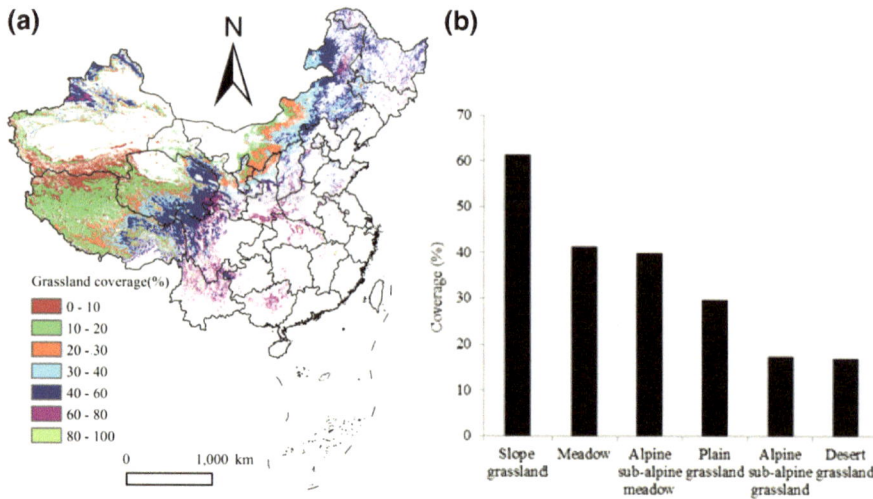

Fig. 2.3 Spatial distribution of the mean grass coverage from 1982 to 2010 in China (**a**) and statistical analysis (**b**)

29.8%, were located in Hulun Buir region, center region of IM and in the south of Ningxia Hui Autonomous Region with coverage ranging from 20 to 40%. In alpine and sub-alpine grasslands, the mean grass coverage was 17.5%. These grasslands mainly distributed in the west of Tibet Autonomous Region and in the west of Qinghai Province with coverage between 10 and 30%. In desert grasslands, the mean grass coverage was 17.1%. These grasslands were located in the middle and west of IM and in the Kunlun Mountains with grass coverage less than 20%.

2.3.2 Temporal Changes of Grass Coverage in China

The change tendency of annual grass coverage was calculated based on Eq. 4 from 1982 to 2010. We found that the total grass coverage increased by 0.17% per year (Fig. 2.4a). The most significant increases were found in the slope grasslands (0.27% per year), the alpine and sub-alpine meadows and the alpine and sub-alpine grasslands (0.174, 0.17% per year); and desert grasslands (0.12% per year). In the plain grasslands and meadows, the lowest increase was observed (0.11% per year, 0.10% per year, respectively; Fig. 2.4a; Table 2.2).

Figure 2.5b is the spatial distribution of significance test of grass cover change. Regions with ESI trend of the grass coverage were mainly distributed in the Mu Us Sandy Land, the Kunlun Mountains, western Tibet, western Xinjiang Uygur Autonomous Region, and the Tianshan Mountains during 1982–2010. Areas with SI trend of the grass coverage were mainly located in the central part of the Tibetan Plateau and in the middle part of the Hexi Corridor (Gansu Province). Areas with

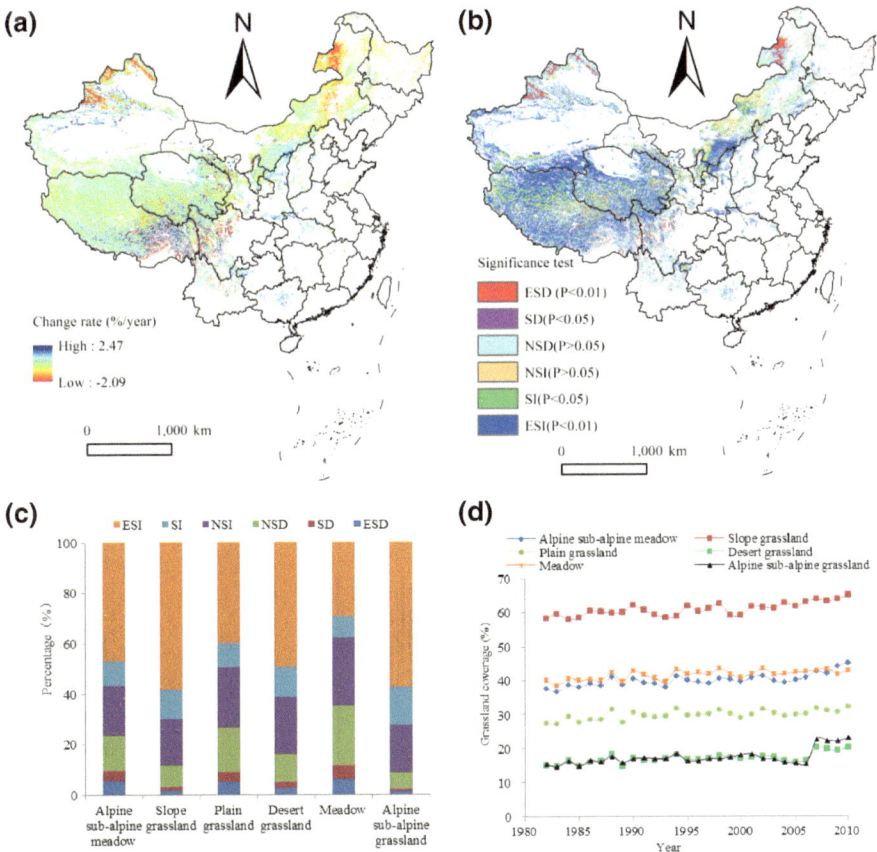

Fig. 2.4 Dynamic of the China's grassland coverage from 1982 to 2010. Pattern of the grass coverage change (**a**), the grade percentage of significant change of different types of grass coverage (**b**), spatial distribution of significant test in China (**c**), and the nationwide inter-annual change of grass coverage (**d**)

ESD trend of the grass coverage were mainly located in Hulun Buir, Tianshan, and Altai Mountains, as well as in the southeast of the Tibetan Plateau. Regions with SD trend of grass coverage were restricted in the Horqin Sandy Land.

For each grassland type, Fig. 2.4c shows the percentage of each grade of significance test as defined in Sect. 2.5. Regions with the ESI and SI trends of grass coverage accounted for 46.03 and 11% of the total grassland area, respectively. Meanwhile, the area with ESD and SD trends accounted only for 4.1 and 3.24% of the total area, respectively. Regions with no significant change accounted for 35.83% (NSI and NSD were 21.83% and 13.8%, respectively). For the slope grasslands, the area percentage with ESI was the largest (58.36%), followed by alpine and sub-alpine grasslands (57.46%). Regions with an increasing trend of grass coverage were larger

Table 2.2 Results of implementation of the grass coverage change' significant test for different grassland types

Percentage (%)	Alpine/sub-alpine meadow	Slope grass-land	Plain grass-land	Desert grass-land	Meadow	Alpine/sub-alpine grassland	Total grass-land
ESD	5.53	1.64	5.26	2.76	6.27	1.22	4.10
SD	4.05	1.68	3.75	2.35	5.30	1.12	3.24
NSD	13.84	8.23	17.61	10.85	23.78	6.47	13.80
NSI	19.90	18.46	23.96	22.69	27.13	18.72	21.83
SI	9.85	11.63	9.55	11.97	8.35	15.02	11.00
ESI	46.83	58.36	39.87	49.39	29.17	57.46	46.03

Note ESD is the abbreviation of extremely significant decrease, SD is significant decrease, NSD is non-significant decrease, NSI is non-significant increase, SI is significant increase, and ESI is extremely significant increase

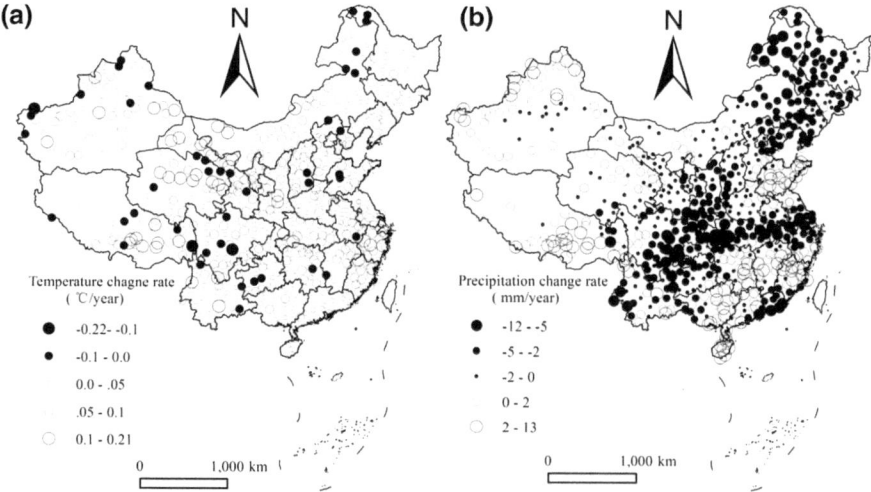

Fig. 2.5 Change trends of annual mean temperature and total precipitation in China from 1982 to 2010

than that showing decreasing trend for all grassland types (Fig. 2.4c). Overall, the grassland coverage had a fluctuation increase trend in this period (Fig. 2.4d).

2.3.3   Correlation Analysis of Grass Coverage and Temperature, Precipitation

2.3.3.1   The Correlation Between Grass Coverage and Temperature, Precipitation

During 1982–2010, temperature increased and precipitation decreased in most parts of China (Fig. 2.5). 84% of the total grassland area showed increasing temperature with a mean rate of 0.04 °C per year. In addition, precipitation decreased in over 56% of the total grassland area with a mean decrease rate of 0.39 mm per year. Precipitation significantly decreased in the east of IM, northeast China, and in the Shanxi, Shaanxi, Sichuan, and Yunnan provinces. In the arid regions of northwest China, such as Xinjiang, Tibet, Qinghai, central part of IM, temperature increased but precipitation is also increased, and the resulting effect of this type of climatic change on the water condition (the sign of the precipitation minus evapotranspiration differences) is unclear. The correlation coefficients of grass coverage with temperature and precipitation were not significant ($R = 0.21$, $P > 0.05$; $R = 0.10$, $P > 0.05$, respectively). The area percentage of the regions where grass coverage positively correlated with precipitation was 78.45% (13.08% with $0.01 < P < 0.05$ and 18.2% with $P < 0.01$). The area percentage of the regions where grass coverage positively correlated with temperature was 62.5% (9.2% with $0.01 < P < 0.05$ and 6.6% with $P < 0.01$). This implies that, in China, the correlations of grass coverage with precipitation are higher than with temperature (Fig. 2.6).

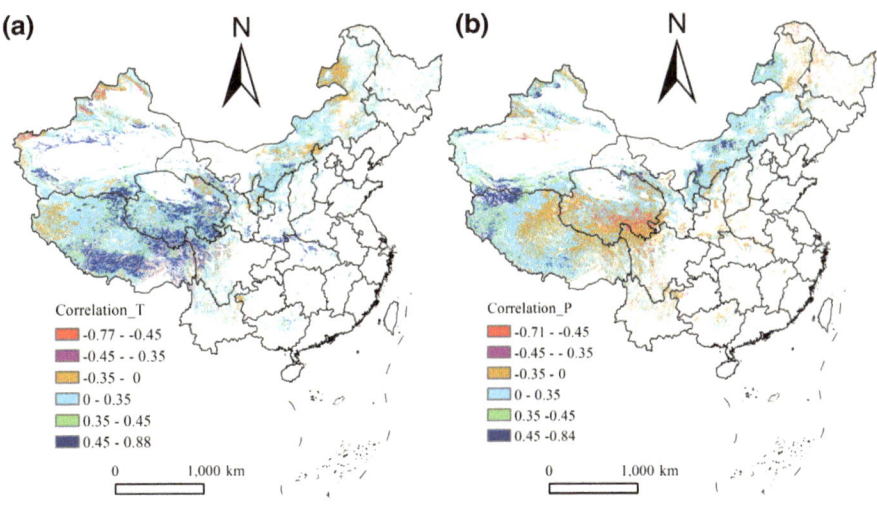

Fig. 2.6 Correlations of grass coverage in China with temperature (**a**) and precipitation (**b**) from 1982 to 2010

The responses of grass coverage to temperature (Fig. 2.6a) and precipitation (Fig. 2.6b) have apparent spatial heterogeneities. The temperature of central and western parts of Tibetan Plateau, south of the Kunlun Mountains, and south of the Tianshan Mountains is the main restrictive factor for grass growth. Here, correlations between grass coverage and precipitation were above 0.35 ($P < 0.05$), while temperature increased during 1982–2010 (Fig. 2.5a). We found significant negative correlations ($P < 0.05$) between grass coverage and temperature in western Xinjiang and in the Hengduan Mountains (Fig. 2.6a). These regions with high temperature, drought, and little rain (Fig. 2.5), especially the Hengduan Mountains, belonged to the dry-hot valleys regions, and precipitation is a limiting factor for vegetation growth. However, we observed significant positive correlations ($P < 0.05$) between grass growth and precipitation in the Kunlun Mountains, western Tibet, and in the northeastern and central parts of IM (Fig. 2.6b). In these regions with desert grassland and little rain, precipitation was a restrictive factor for grass growth and it increased during 1982–2010 (Fig. 2.5b). In the regions with abundant precipitation, such as the Three Rivers Source, central and southern parts of Tibet, we found negative correlations between grass coverage and precipitation. Possible reasons for these negative correlations may be that in the above-mentioned regions with alpine and sub-alpine meadow, the moisture conditions were good but low temperatures restricted the grass growth. Besides, in the regions of Tianshan Mountains, Altai Mountains, negative correlations between grass coverage and precipitation are due to the fact that these are high altitudinal regions with abundant precipitation (and precipitation still increased during 1982–2010). Therefore, here temperature is a main limiting factor for grass growth.

2.3.3.2 The Coupling Relationship Analysis Between Coverage and Temperature, and Precipitation in Different Grassland Types

The correlation between vegetation coverage and temperature derived from 220 meteorological stations in grassland distribution regions represented the coordinate axis X, and coordinate axis Y was the correlation between vegetation coverage and precipitation as showed in Fig. 2.7. The distribution pattern of scatters was different for the six grassland types. For alpine and sub-alpine meadow, scatters were located in the first, second, and fourth quartiles (Fig. 2.7a), and the correlation between coverage and temperature, precipitation was 0.24, 0.005, respectively. It indicated that grass growth was more affected by temperature. Slope grassland and scatters were distributed in the first and fourth quartiles (Fig. 2.7b), which reflected the distribution characteristics. The correlation between coverage and temperature and precipitation was 0.28 and 0.01, respectively. Plain grassland scatters, located in first and second quartiles, are mainly distributed in the central of IM, east of Ningxia and north of Xinjiang with low precipitation. The correlation between coverage and temperature and precipitation was 0.14, 0.16, respectively. Desert grassland scatters were located in the first and fourth quartiles. The correlation between coverage and temperature

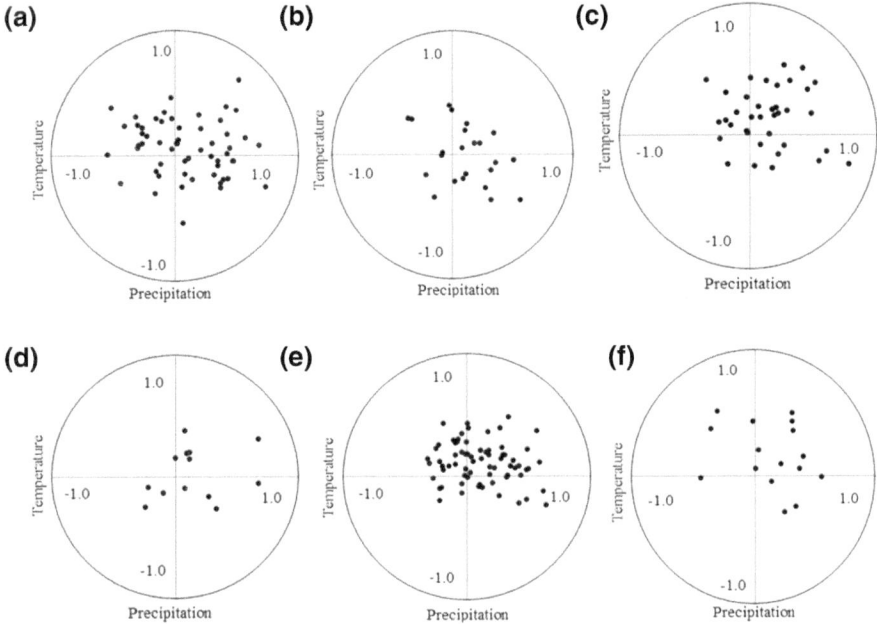

Fig. 2.7 Distribution of correlation coefficient of temperature, precipitation, and grassland coverage of different grassland types (**a** denotes alpine and sub-alpine meadow, **b** denotes slope grassland, **c** denotes plain grassland, **d** denotes desert grassland, **e** denotes meadow, and **f** denotes alpine and sub-alpine grassland)

and precipitation was 0.18 and 0.23, respectively. Meadow scatters, located in the first, second and fourth quartiles, reflected the distribution characteristic of grassland. The correlation between coverage and temperature and precipitation was 0.15 and 0.08, respectively. Alpine and sub-alpine grassland scatters were located in the first and second quartiles, and the correlation between coverage and temperature and precipitation was 0.24 and 0.12, respectively. In conclusion, grass coverage in plain grasslands and desert grasslands were more affected by precipitation than temperature; and the rest of four grassland types and grass coverage were more affected by temperature.

2.3.3.3 Correlations of Monthly Grass Coverage and Climate Factors; Time-Lag Effect

Monthly grass coverage, temperature, and precipitation data during 1982–2010 were used to analyze correlations of the current month's grass coverage and the zero to three former months' temperature and precipitation (Figs. 2.8 and 2.9). The mean correlation coefficient between the current month's grass coverage and current month's temperature was 0.8 (Fig. 2.8a), which was much higher than that on inter-annual scale (0.21). Statistical analysis showed that the area percentage of the regions with

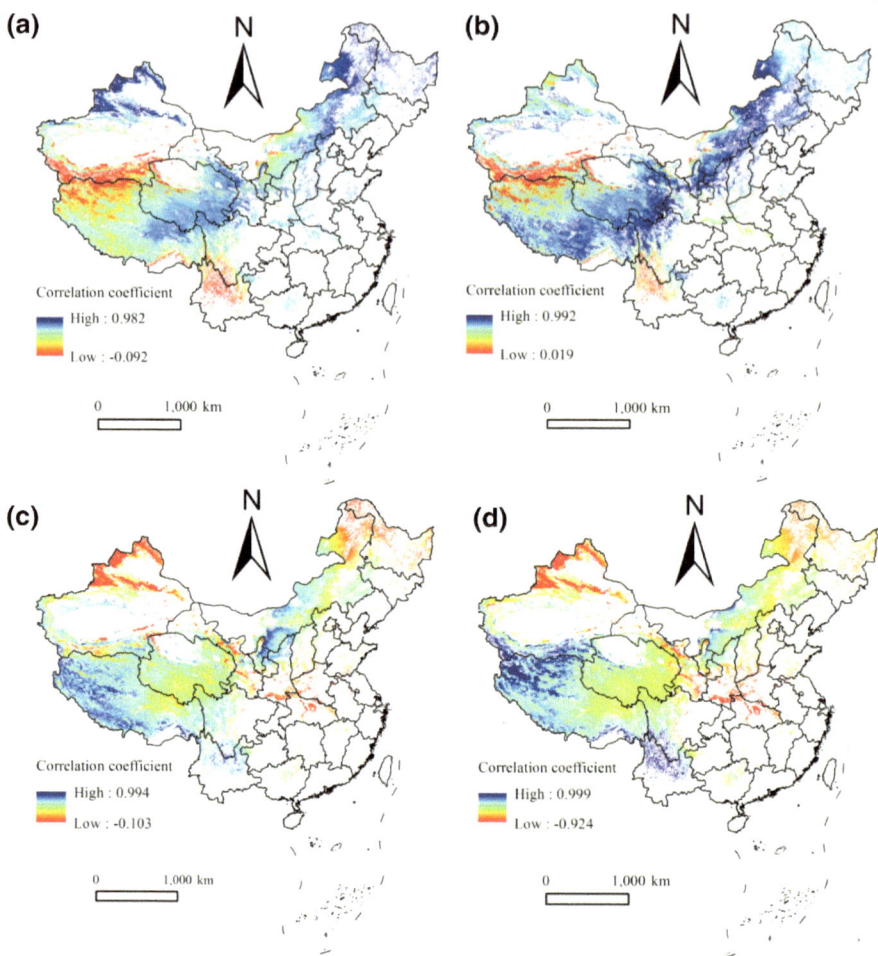

Fig. 2.8 Correlation coefficients between current month's grass coverage and the current month's temperature (**a**), the former one month's temperature (**b**), the former two months' temperature (**c**), and the former three months' temperature (**d**)

grass coverage positively correlated with temperature was 98% (1%, with $0.01 < P < 0.05$ and 96.3% with $P < 0.01$). This indicated that the grass growth was strongly affected by monthly temperature variations. Furthermore, correlation between current month's grass coverage and the former one month's temperature was the largest (0.90), followed by correlations with the former two months' temperatures (0.74). Area percentage of the regions where vegetation coverage was positively correlated with former one to three months' precipitation was 99.5, 98.0, and 85.60%. These correlations showed an obvious time-lag effect when the grass growth was more affected by the former one month's temperature than by the current month's temperature.

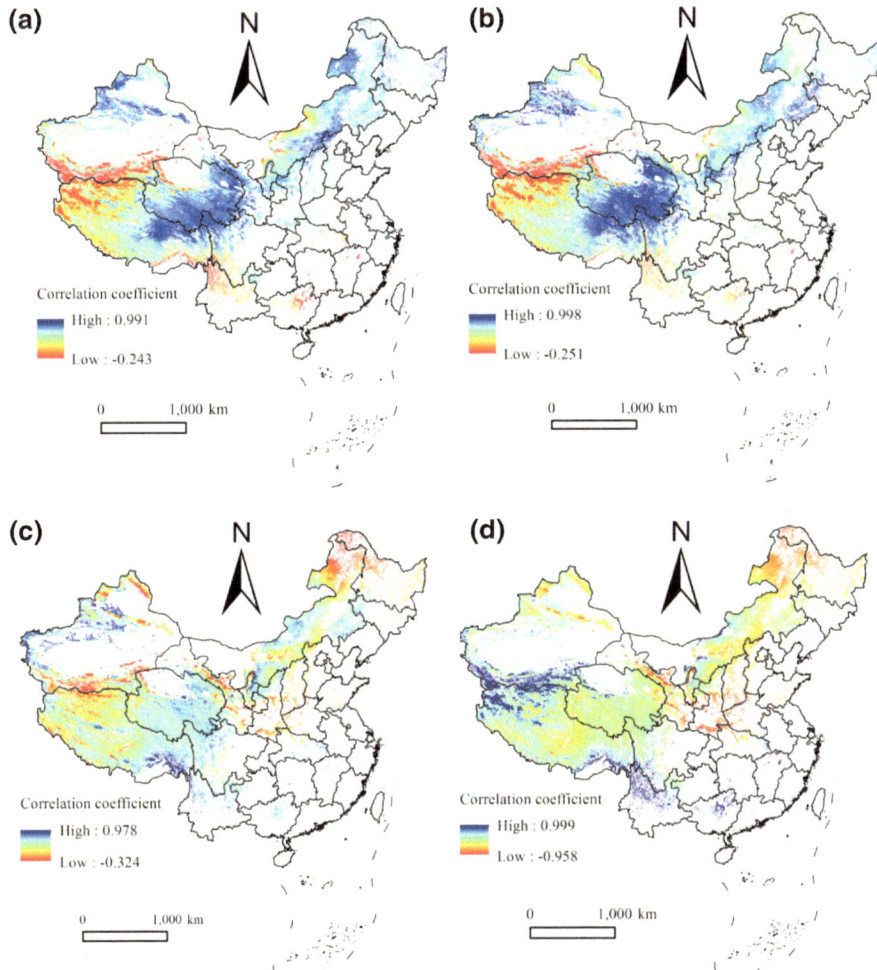

Fig. 2.9 Correlation coefficients between current month's grass coverage and the current month's precipitation (**a**), the former one month's precipitation (**b**), the former two months' precipitation (**c**), and the former three months' precipitation (**d**)

The mean correlation coefficient between the grass coverage and current month's precipitation was 0.76 (Fig. 2.8a), much higher than that on the inter-annual time scale (0.10). Area percentage of the regions with grass coverage positively correlated with precipitation was 98% (3.6% with $0.01 < P < 0.05$ and 93.6% with $P < 0.01$). Correlation between grass coverage and the former one month's precipitation was also the highest (Fig. 2.9b). The correlations between the current month's grass coverage and former one to three months' precipitation equal to 0.79, 0.56, and 0.03, respectively. Area percentage of the regions where vegetation coverage was positively correlated with former one to three months' precipitation was 99.5%,

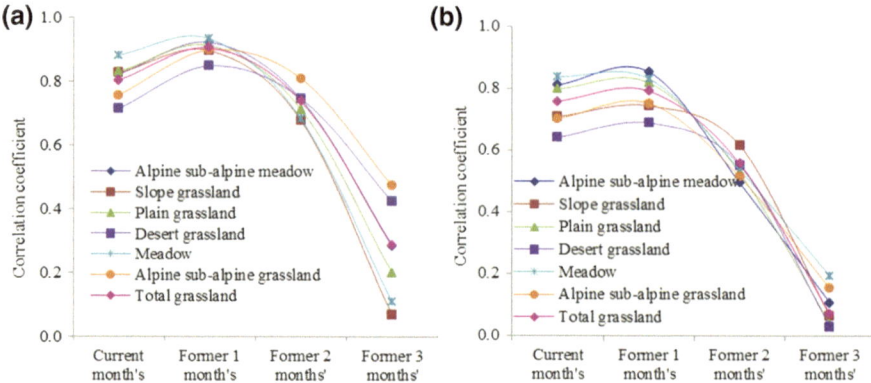

Fig. 2.10 Correlation coefficients between current months' grass coverage for different grassland types and the current month's, former one month's, former two months', and former three months' temperature (**a**) and precipitation (**b**)

99%, and 45.57%, respectively. Therefore, grass coverage also had one-month time-lag with precipitation.

2.3.3.4 Response of Grass Coverage in Different Grassland Types to Climatic Variations

Responses of grass coverage for different grassland types to temperature and precipitation are different; overall, correlations were highest with the former one month's temperature and precipitation (Fig. 2.10). For alpine and sub-alpine meadow, the correlation of the current month's grass coverage with the former zero to three months' temperature is 0.82, 0.92, 0.75, and 0.28, respectively, and with precipitation are 0.81, 0.85, 0.61, and 0.06. Thus, for these grasslands mostly distributed across the Tibetan Plateau, both temperature and precipitation were responsible for the grass growth which was more affected by the former one month's temperature and precipitation (time-lag effect). For all the other five grassland types, the correlation of current month's grass coverage with former one month's temperature and precipitation was largest, and the correlation between coverage and temperature was larger than that with precipitation as shown in Tables 2.3 and 2.4. Apparently, grass growth has one-month time-lag effect to temperature and precipitation.

2.3.4 Comparisons

Climate change has important effects on terrestrial vegetation. In the last two decades, vegetation coverage in most region of China gradually increased. In northwest China and the Tibetan Plateau, both temperature and precipitation showed increasing trends

Table 2.3 Correlation coefficients between current month's grass coverage for different grassland types and the current month's, former one month's, former two months', and former three months' temperature

Temperature	Current month's	Former one month's	Former two months'	Former three months'
Alpine/sub-alpine meadow	0.82	0.92	0.74	0.28
Slope grassland	0.83	0.90	0.68	0.07
Plain grassland	0.83	0.91	0.71	0.20
Desert grassland	0.71	0.85	0.75	0.42
Meadow	0.88	0.93	0.68	0.11
Alpine/sub-alpine grassland	0.76	0.90	0.81	0.47
Total grassland	0.80	0.90	0.74	0.29

Table 2.4 Correlation coefficients between current month's grass coverage for different grassland types and the current month's, former one month's, former two months', and former three months' precipitation

Precipitation	Current month's	Former one month's	Former two months'	Former three months'
Alpine/sub-alpine meadow	0.81	0.85	0.49	0.11
Slope grassland	0.71	0.74	0.61	0.06
Plain grassland	0.80	0.82	0.53	0.04
Desert grassland	0.64	0.69	0.55	0.03
Meadow	0.84	0.83	0.54	0.19
Alpine/sub-alpine grassland	0.70	0.75	0.51	0.15
Total grassland	0.76	0.79	0.56	0.07

in the past 29 years that was consistent with Shi et al. (2007), who concluded that the pattern of climate change in northwest China became warmer and more humid since 1980s. In support of Fang et al. (2003), our findings found that the grassland coverage showed an increase trend from 1982 to 2010 in 78.9% of total grassland area of China and the areas of extremely significant and significant increases mainly distributed in the above-mentioned regions with warmer and wetter climate conditions.

In this study, there were no significant positive correlations between grass coverage with temperature and precipitation on the inter-annual time scale, while on the monthly time scale, these correlations were significant. The area percentage of regions where correlations of grass coverage and temperature, precipitation reached significant positive level (i.e., R-values were statistically significant at the 0.05 level) was 98% and 97.2%, respectively. These findings were consistent with a study by Dai (2010), who analyzed northwestern China's vegetation cover and concluded that on the monthly time scale, correlations between grass growth and climate factors were significant. Many studies concluded that the grass growth has significant correlation with precipitation (Li et al. 2006; Piao et al. 2006; Sun et al. 2013). Our study concluded that the correlation between grass growth and temperature and precipitation was not significant at the annual time scale, but only on the monthly time scale correlation was significant. Moreover, in the most of grassland types (4 grasslands of 6), the grass growth variations are more affected by temperature than by precipitation.

However, grass growth is affected not only by climate conditions but is also influenced by human activities, especially with human intervention intensity being increasing at present. In the past decade, the implementation of ecological restoration programs, such as the "Grain to Green" program since 1999 and "Returning of Grazing Land to Grassland" since 2003 have been leading to the restoration of grassland. This interventions decrease the sensitivity of grass growth to climate change to some extent, reducing correlation coefficient between grass coverage and climate factors. Zhang et al. (2011) analyzed spatial and temporal variations of vegetation coverage in the Loess Plateau during 1982–2009 and concluded that during 1982–1998, the annual mean NDVI showed strong correlations with temperature and precipitation. However, during 1999–2009, the annual mean NDVI increased significantly following the implementation of large-scale ecological restoration projects and the sensitivity of vegetation coverage to climate factors decreased. Mu et al. (2013) also reported there were weak correlations between vegetation coverage and climate factors on the inter-annual time scale. It looks like the vegetation coverage variations are a result of the combined actions of climate change and human activities (Xin et al. 2008). At some extent, climate change plays a dominant role in global vegetation coverage variations, while human activities often execute decisive impacts on spatial–temporal change of vegetation (Chen et al. 2000). For example, human activities have a significant role in promoting grass growth and restoration (e.g., forbidding grazing and rotation grazing were beneficial to grassland restoration in Maqu County (Wang et al. 2009; Ta et al. 2008) concluded that grass coverage increased by 97% after four years forbidding grazing in Alxa desert grassland, etc.). Besides, vegetation coverage significantly increased in Korqin Sandy Land after the implementation of ecological restoration program (Zhang et al. 2012). And the increase of grass net primary productivity was mostly attributed to ecological restoration programs (Zhou et al. 2013). These examples confirmed the anthropogenic role in the spatial and temporal variations of grass growth.

2.4 Conclusions

GIMMS NDVI and MODIS NDVI were used to calculate grass coverage from 1982 to 2010 and analyze its variations and correlation with major climatic variables (temperature and precipitation) at inter-annual and monthly time scales. The following conclusions can be drawn:

(1) The spatial distribution of grassland vegetation coverage in China had obvious heterogeneity with the grass coverage being higher in southeast and lower in northwest China. For the whole region, the mean grass coverage was 34% during 1982–2010. The grass coverage was the highest in slope grassland, while the lowest in desert grassland.

(2) For 78.9% of the total grassland area of China, its grass coverage is increasing, and the mean coverage increased by 0.17% per year. The increase rates of grass coverage during the past 29 years for different grassland types can be sorted in the declining order as follows: slope grassland, alpine and sub-alpine meadow, alpine and sub-alpine grassland, desert grassland, plain grassland, and meadow. The slope grassland' grass coverage has the largest increase rate equal to 0.27% per year and lowest rates in plain grassland and meadow (0.11%/year and 0.1%/per year, respectively).

(3) Area percentage of grass coverage with extremely significant and significant increase trend were 46.03% and 11%, respectively; while the area percentage of grass coverage with extremely significant and significant decreases were only 4.1 and 3.24%.

(4) On the inter-annual time scale, correlations of grass coverage with temperature and precipitation were not significant. The increases of both temperature and precipitation promote grass growth, but the responses of the grass coverage to climatic factors for different grassland types are different. The grass growth was more affected by temperature than by precipitation in alpine and sub-alpine grassland, alpine and sub-alpine meadow, slope grassland and meadow, while grass growth was more affected by precipitation in desert grassland and plain grassland.

(5) On the monthly time scale, positive correlations of the grass coverage with temperature and precipitation were larger than that on the inter-annual scale and were statistically significant correlation. This indicates that seasonal fluctuations of the hydrothermal conditions have larger impacts on the grass growth than the inter-annual fluctuations. Overall, grass coverage was more affected by temperature than by precipitation (in all the six grassland types). There was one-month time-lag effect and the correlations were largest between the current month grass coverage and the former one month's temperature and precipitation.

References

Chen X, Tan Z, Schwartz MD, Xu C (2000) Determining the growing season of land vegetation on the basis of plant phenology and satellite data in northern China. Int J Biometeorol 44(2):97–101

Christensen L, Coughenour MB, Ellis JE, Chen ZZ (2004) Vulnerability of the Asian typical steppe to grazing and climate change. Clim Change 63(3):351–368

Cui X, Graf H (2009) Recent land cover changes on the Tibetan Plateau: a review. Clim Change 94(1–2):47–61

Dai Z (2010) Intensive agropastoralism: dryland degradation, the Grain-to-Green Program and islands of sustainability in the Mu Us Sandy Land of China. Agr Ecosyst Environ 138(3):249–256

Fang JY, Piao SL, Tang Z, Peng C, Ji W (2001) Interannual variability in net primary production and precipitation. Science 293(5536):1723

Gan CY, Wang XZ, Bao-Sheng LI, Liang ZX, Zhi-Wen LI, Wen XH (2011) Changes of vegetation coverage during recent 18 years in Lianjiang River watershed. Sci Geog Sin 31(8):1019–1024

GLC (2003) Global land cover classification for the year 2000. http://www-gem.jrc.it/glc2000/

Gutman G, Ignatov A (1998) The derivation of the green vegetation fraction from NOAA/AVHRR data for use in numerical weather prediction models. Int J Remote Sens 19(8):1533–1543

Hall FG, Huemmrich KF, Goetz SJ, Sellers PJ, Nickeson JE (1992) Satellite remote sensing of surface energy balance: success, failures, and unresolved issues in FIFE. J Geophys Res: Atmos (1984–2012), 97(D17):19061–19089

Keeling CD, Chin J, Whorf TP (1996) Increased activity of northern vegetation inferred from atmospheric CO_2 measurements. Nature 382(6587):146–149

Li SG, Harazono Y, Oikawa T, Zhao HL, Ying He Z, Chang XL (2000) Grassland desertification by grazing and the resulting micrometeorological changes in inner Mongolia. Agric For Meteorol 102(2):125–137

Li CZ, Ma MG, Zhang F, Jiang ZR (2006) The dynamic analysis of vegetation pattern in the northwest of China. Remote Sens Technol Appl 21(4):332–337

Luo L, Wang ZM, Song KS, Zhang B, Liu DW, Ren CY (2009) Research on the correlation between NDVI and climatic factors of different vegetations in the northeast China. Acta Botanica Boreali-Occidentalia Sin 4:800–808

Mu SJ, Yang HF, Li JL, Chen YZ, Gang CC, Zhou W (2013) Spatio-temporal dynamics of vegetation coverage and its relationship with climate factors in inner Mongolia, China. J Geog Sci 23(2):231–246

Myneni RB, Keeling CD, Tucker CJ, Asrar G, Nemani RR (1997) Increased plant growth in the northern high latitudes from 1981 to 1991. Nature 386(6626):698–702

Parmesan C, Yohe G (2003) A globally coherent fingerprint of climate change impacts across natural systems. Nature 421(6918):37–42

Piao SL, Guo QH (2001) Application of CASA model to the estimation of Chinese terrestrial net primary productivity. Acta Phytoecologica Sin 25(5):603–608

Piao SL, Fang JY, He JS (2006) Variations in vegetation net primary production in the Qinghai-Xizang Plateau, China, from 1982 to 1999. Clim Change 74(1):253–267

Ran Y, Li X, Lu L (2010) Evaluation of four remote sensing based land cover products over China. Int J Remote Sens 31(2):391–401

Ren JZ, Liang TG, Lin HL, Feng QS, Huang XD, Hou FJ (2011) Study on grassland's responses to global climate change and its carbon sequestration potentials. Acta Prataculturae Sin 20(2):1–22

Roerink GJ, Menenti M, Soepboer W, Su Z (2003) Assessment of climate impact on vegetation dynamics by using remote sensing. Phys Chem Earth, Parts A/B/C 28(1):103–109

Shi YF, Shen YP, Kang E, Li DL, Ding YJ, Zhang G (2007) Recent and future climate change in northwest China. Clim Change 80(3):379–393

Sun H, Wang C, Zheng N, Bukhosor (1998) Analysis of the vegetation cover change and the relationship between NDVI and environmental factors by using NOAA time series data. J Remote Sens 266(2):153–161

Sun J, Cheng GW, Li WP, Sha YK, Yang YC (2013) On the variation of NDVI with the principal climatic elements in the Tibetan Plateau. Remote Sens 5(4):1894–1911

Ta L, TL, CJ, Li J, Zhang JW (2008) Monitoring on the effect of returning grazing desert to grassland in Alashan. Pratacyltural Sci 25(2):124–127

Tang H (2003) Intra-annual variability of NDVI and its relation to climate in northeast China transect. Quat Sci 23(3):318–325

Walker B, Steffen W (1997) IGBP science no. 1: a synthesis of GCTE and related research. IGBP, Stockholm, pp 1–24

Wang J, Guo N, Cai DH, Deng ZY (2009) The effect evaluation of the program of restoring grazing to grasslands in Maqu County. Acta Ecol Sin 29(3):1276–1284

Weltzin JF, Loik ME, Schwinning S, Williams DG, Fay PA, Haddad BM (2003) Assessing the response of terrestrial ecosystems to potential changes in precipitation. Bioscience 53(10):941–952

Wu DD, Cai YL (2009) Evaluation of ecological restoration effects in China: a review. Prog Geogr 28(4):622–628

Xin ZB, Xu JX, Zheng W (2008) Spatiotemporal variations of vegetation cover on the Chinese Loess Plateau (1981–2006): impacts of climate changes and human activities. Sci China, Ser D Earth Sci 51(1):67–78

Zhang WH, Yang W (2011) The feature analysis for grassland degradation and the restoration of natural vegetation in degraded grassland. Northern Environ (8):40–44

Zhang W, Zhang H, Ze B (2006) Progress studies on the carbon cycle of Alpine meadow in China. J Mt Sci 24(B10):266–274

Zhang GL, Dong JW, Xiao XM, Hu ZM, Sheldon S (2012) Effectiveness of ecological restoration projects in Horqin Sandy Land, China based on SPOT-VGT NDVI data. Ecol Eng 38(1):20–29

Zhao MS, Fu ZB, Yan XD, Wen G (2001) Study on the relationship between different ecosystems and climate in china using NOAA/AVHRR data. Acta Geogr Sin 56(3):287–296

Zhou LM, Tucker CJ, Kaufmann RK, Slayback D, Shabanov NV, Myneni R (2001) Variations in northern vegetation activity inferred from satellite data of vegetation index during 1981 to 1999. J Geophys Res: Atmos (1984–2012), 106(D17):20069–20083

Zhou W, Li JL, Mu SJ, Gang C, Sun ZG (2013) Effects of ecological restoration-induced land-use change and improved management on grassland net primary productivity in the Shiyanghe River Basin, north-west China. Grass Forage Sci 69:596–610

Chapter 3
Spatial–Temporal Dynamics of Grassland Net Primary Productivity in China and Its Response to Climate Factors

Abstract Grasslands in China are mainly located in ecologically fragile regions, which are sensitive to climate change. During recent decades, grasslands have experienced prominent warming and drying. Therefore, investigating the response of grasslands to climatic variations is necessary for a better understanding of the cumulative consequences of climate change. Grassland net primary productivity (NPP) is an important indicator for evaluating grassland ecosystem conditions. We used multisource remote sensing data and meteorological data to estimate the grassland NPP from 1982 to 2010, based on the Carnegie–Ames–Stanford approach (CASA) model. The spatial pattern and change trends of grassland NPP were evaluated. The response of grassland NPP changes to climatic variations was also analyzed. The results revealed that the spatial distribution of grassland NPP showed an increasing trend from the northwest to southeast across China. During the period from 1982 to 2010, the grassland mean NPP was 282 gC/m^2/year, and NPP exhibited apparent spatial heterogeneity, being highest (710 gC/m^2/year) in the dry heat savanna shrub and grass and lowest (58 gC/m^2/year) in the alpine desert. The total annual NPP was 988.3 Tg C. Grassland NPP has increased slightly in the past 30 years, at a rate of 0.6 gC/m^2/year. Regions showing increasing NPP accounted for 67.2% of the total grassland areas, within which areas with extremely significant ($P < 0.01$) and significant ($P < 0.05$) increases accounted for 35.8% and 8.0% of the total grassland area, respectively. Regions showing extremely significant and significant decreases encompassed only 5.8% and 4.8% of total grassland area, respectively. Clear increases in grassland NPP were observed in the west of the Qinghai–Tibet Plateau, the Alxa Plateau, and western area of Xinjiang. Areas with a decrease in grassland NPP were mainly distributed in the western regions of Inner Mongolia. Furthermore, the increasing rate of NPP showed temporal variation and differed among different grassland types. The correlation coefficient between NPP and precipitation was larger than that between NPP and temperature. Moreover, the response of grassland NPP to temperature and precipitation differed for different grassland types. There were significant positive correlations between annual precipitation and NPP in temperate desert steppe, temperate steppe, and temperate meadow steppe.

Keywords Grassland net primary productivity · Spatial–temporal dynamics · Temperature · Precipitation · The coupling relation analysis

© Springer Nature Singapore Pte Ltd. 2020
W. Zhou et al., *Remote Sensing Monitoring and Evaluation of Degraded Grassland in China*, Springer Geography,
https://doi.org/10.1007/978-981-32-9382-3_3

3.1 Introduction

The response of terrestrial ecosystems to climate change is one of the focuses of global change research (Walker and Steffen 1997). Vegetation is the most intuitive manifestation and important symbol of climate change. The study of NPP in terrestrial ecosystems at global or regional scales began in the mid-nineteenth century. There are different opinions on the impact of climate factors on NPP in terrestrial ecosystems in academic circles (Fang et al. 2001). Studies have shown that NPP of global terrestrial vegetation increased from 1980 to 2000. However, the reasons for the increase are different in different regions (Myneni et al. 1997). The coupling relationship between global terrestrial vegetation NPP and climate data from 1982 to 1999 has also been comprehensively analyzed. It is found that climate change alleviates climate stress and increases global terrestrial vegetation NPP by 6% (Running et al. 2004). Under the background of global climate change, China's climate has changed correspondingly (Shi et al. 2007; Zhai et al. 2005), and NPP of terrestrial vegetation shows a certain growth trend (Piao et al. 2001). Climate change weakens the stress of hydrothermal conditions on vegetation growth, and the response of vegetation NPP to climate change in China has obvious regional differences (Zhu et al. 2006). Grassland, as an important part of terrestrial ecosystem, covers 20% of the land surface. Because of its huge coverage area and diversity of types, grassland has a unique position and importance in the study of regional and global carbon cycle (Ni 2004; Scurlock and Hall 1998). In China, grasslands have an area of approximately 4 million km^2, accounting for 41.7% of the national land area of China, and mainly distributed in arid and semi-arid climates of the north and alpine climatic zones in the Tibetan Plateau. The ecological environment of grasslands is relatively fragile and sensitive to climate change (Christensen et al. 2004). NPP is an important characterization of carbon sequestration capacity of grassland ecosystem. Therefore, quantitative study on the temporal and spatial characteristics of grassland NPP and its response to climate change in China can provide a scientific basis for correctly evaluating grassland ecosystem productivity and help to understand the mechanism of global change affecting carbon cycle of terrestrial ecosystem. In recent years, domestic and foreign scholars have carried out extensive research on grassland NPP simulation and its response to climate change in China (Piao et al. 2007; Zhang et al. 2008). Because of the differences of time and spatial scales and data sources, the comparability of NPP simulation results is poor, and the responses of grassland NPP to climate, water, and heat factors are different in different regions.

In view of the above, in order to better understand and comprehensively analyze the temporal and spatial dynamics of grassland NPP and its relationship with climate change in China in long time series, the following studies were carried out in China grassland ecosystem from 1982 to 2010. (1) Based on CASA model, using NDVI data, meteorological data, and land-cover data, the dynamic simulation of grassland NPP in China was carried out to clarify the spatial and temporal characteristics of grassland NPP in the past 30 years. (2) Through the coupling analysis of grassland NPP and meteorological data in China, the change trend of grassland NPP in different periods and types and its response characteristics to climate, water, and heat factors were revealed.

It has important theoretical and practical significance for the study of carbon cycle and carbon budget of grassland ecosystem under the background of global climate change and can provide scientific basis for the protection of grassland resources and the sustainable development of grassland ecosystem.

3.2 Data Source and Processing

3.2.1 GIMMS NDVI Data and Processing

In this study, the largest half-month composite data of NDVI from the National Aeronautics and Space Administration (NASA) and Global Monitoring and Modeling Research Group (GIMMS) were used as the data source (http://ecocast.arc.nasa.gov/). The time series was from 1982 to 2010, and the spatial resolution is 8×8 km. The dataset was the longest time series of NDVI data. GIMMS NDVI dataset with small error and high precision has been widely used in global and regional large-scale vegetation change research. Because Max Value Composite (MVC) can eliminate some clouds, atmosphere, and solar elevation angle interference, this paper used MVC to obtain monthly NDVI data, used WGS-1984 geographical coordinate system and Albers Equal-Area Conic projection.

3.2.2 Meteorological Data and Processing

Meteorological data from 1982 to 2010, including average monthly temperature and precipitation for 720 stations as well as total solar radiation data for 102 stations, were obtained from China Meteorological Data Sharing Service System. According to the longitude, latitude, and elevation information of each meteorological station, the ANUSPLIN software was used to spatially interpolate the meteorological data to obtain the meteorological raster image as same as the resolution and projection system of NDVI data.

3.2.3 Grassland-Type Data

The grassland classification data were derived from the national grassland resource survey in the 1980s (China 1996). According to the principle of vegetation habitat (topography, soil, and climate), the grassland in China was divided into 19 categories, including temperate meadow steppe, temperate steppe, temperate desert

steppe, alpine meadow steppe, alpine steppe, alpine desert steppe, temperate steppification desert, temperate desert, alpine desert, warm grass, warm shrub grass, hot
grass, hot shrub, dry savanna shrub, lowland meadow, mountain meadow, alpine
meadow, swamp, and improved grassland. The classification system had a grassland
area of 3.52 million km^2, accounting for 37% of China's land area.

3.3 Methodology

3.3.1 CASA Model

Grassland NPP was calculated using the CASA model, which is a light-use efficiency
model based on the resource-balance theory (Potter et al. 1993; Field et al. 1995). In
the CASA model, NPP is the product of absorbed photosynthetically active radiation
(APAR) and light-use efficiency (ε). The basic principle of the model can be described
by the following formula:

$$NPP(x, t) = APAR(x, t) \times \varepsilon(x, t) \tag{3.1}$$

where x is the spatial location (pixel number), t is time, APAR is the canopy-absorbed
incident solar radiation integrated over a given time period (MJ/m^2), and $\varepsilon(x, t)$ is the
actual light-use efficiency(gC/MJ). APAR(x, t) is calculated based on the following
formula:

$$APAR(x, t) = SOL(x, t) \times FPAR(x, t) \times 0.5 \tag{3.2}$$

where SOL(x, t) is total solar radiation (MJ m-2) of pixel x in time t, 0.5 is the
proportion of SOL available for vegetation (wavelength range of 0.38–0.71 μm),
and FPAR(x, t) is the fraction of PAR absorbed by vegetation and calculated by the
linear function of NDVI simple ratio (SR) as follows:

$$SR = [1 + NDVI(x, t)] / [1 - NDVI(x, t)] \tag{3.3}$$

$$FPAR(x, t) = \min \left[\frac{SR(x, t) - SR_{min}}{SR_{max} - SR_{min}}, 0.95 \right] \tag{3.4}$$

where SR$_{min}$ is SR for bare land and is set to 1.05 for all grid cells, and SR$_{max}$ is
for land independent of vegetation types and in this study, the SR$_{max}$ for grassland is
4.46 (Piao et al. 2006). A cap of 0.95 was imposed on FPAR to reflect a finite upper
limit to leaf area. The algorithm of $\varepsilon(x, t)$ can be expressed as follows:

$$\varepsilon(x, t) = T_{\varepsilon 1}(x, t) \times T_{\varepsilon 2}(x, t) \times W_{\varepsilon}(x, t) \times \varepsilon_{max} \tag{3.5}$$

Fig. 3.1 Field survey sites and grassland types. *Note* 0 Non-grassland, 1 temperate meadow steppe, 2 temperate steppe, 3 temperate desert steppe, 4 alpine meadow steppe, 5 alpine steppe, 6 alpine desert steppe, 7 temperate steppification desert, 8 temperate·desert, 9 alpine desert, 10 warm grass, 11 warm shrub grass, 12 hot grass, 13 hot shrub, 14 dry savanna shrub, 15 lowland meadow, 16 mountain meadow, 17 alpine meadow, 18 swamp, 19 improved grassland

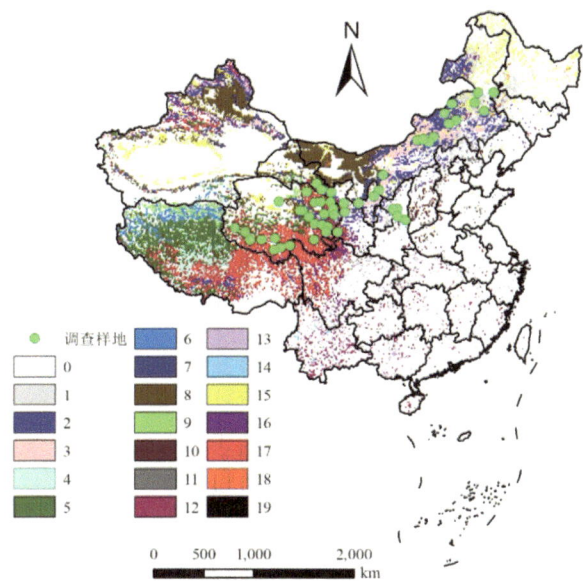

where $T_{\varepsilon 1}(x, t)$ and $T_{\varepsilon 2}(x, t)$ are temperature stress coefficients, $W_\varepsilon(x, t)$ is the water stress coefficient that indicates the reduction of light-use efficiency caused by moisture factor. A more detailed description of $T_{\varepsilon 1}(x, t)$, $T_{\varepsilon 2}(x, t)$, and $W_\varepsilon(x, t)$ can be found in Yu et al. (2011).

ε_{max} indicates the maximal light-use efficiency under ideal conditions. And ε_{max} is easily affected by actual temperature and moisture conditions and differs greatly compared with real conditions (Paruelo et al. 1997). The value of ε_{max} for grassland is 0.542 in this study in accordance with the study of Zhu et al. (2006) (Fig. 3.1).

3.3.2 Validation of CASA Model

Validation was conducted by comparing the field observation data for grassland with the estimated data by CASA model. In order to verify the accuracy of the model, 51 sample plots were set up in July and August 2009 during the season when grassland vegetation was growing vigorously. At each site (10 × 10 m), we set five quadrates (1 × 1 m) with flat terrain and uniform grassland distribution. In each quadrate, the aboveground parts of the plants were harvested evenly and then dried in a constant temperature oven at 70 °C until the dry weight was constant. According to the distribution ratio of underground and aboveground biomass in different grassland types published in China (Piao et al. 2004), the underground biomass was calculated according to the root-shoot ratio and aboveground biomass, and then the carbon conversion coefficient was 0.475, and NPP was obtained. Figure 3.2 presents the results of the correlation between the observed NPP and estimated NPP. The correlation

Fig. 3.2 Consistency test of grassland NPP field measured value and CASA model simulation value

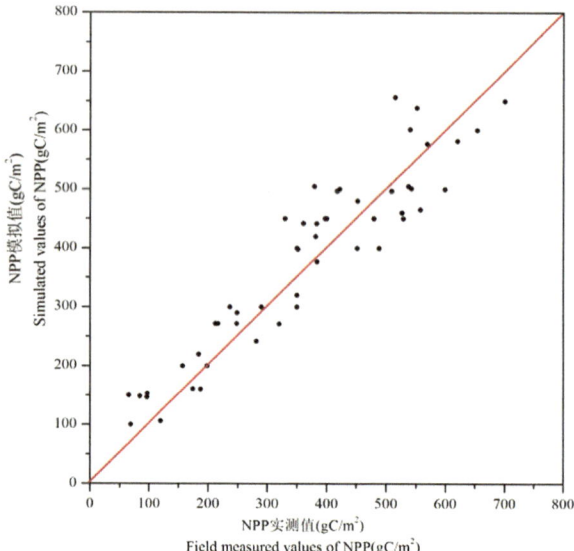

was significance, which indicates that the model's estimation accuracy is reliable in grassland of China. The simulated and measured values of NPP are distributed near the 1:1 line, and the statistical analysis showed that the relative error (REE = (simulated-measured value)/measured value) was between 17 and 30%, with an average accuracy of 86%. It can be seen that the simulation results of the CASA model were ideal and highly consistent with the measured NPP values. The correlation was significant, which indicates that the model's estimation accuracy is reliable in grassland of China.

3.3.3 Inter-annual Variation Trend of Grassland NPP

Vegetation dynamics measured by NPP are the most intuitive manifestation of grassland degradation. Equation 3.4 can be used to calculate the slope of PNPP and LNPP by ordinary least square estimation (Ma and Frank 2006). The formula is expressed as follows:

$$\text{Slope} = \frac{n \times \sum_{i=1}^{n} i \times \text{NPP}_i - \left(\sum_{i=1}^{n} i\right)\left(\sum_{i=1}^{n} \text{NPP}_i\right)}{n \times \sum_{i}^{n} i^2 - \left(\sum_{i=1}^{n} i\right)^2} \tag{3.4}$$

where i is 1 for year 1982, 2 for year 1983, and so on; n is 29 for years 1982 to 2010; and NPP_i is the value of annual NPP in time of i year.

F-test was used to test the significance of variation trend, and the significance only represents the level of confidence of the trend change and has nothing to do with the speed of change. The statistic calculation formula is:

$$F = U \times \frac{n-2}{Q} \tag{3.5}$$

$U = \sum_{i=1}^{n} (\hat{y}_i - \bar{y})^2$ is called the regression square sum, $Q = \sum_{i=1}^{n} (y_i - \hat{y}_i)^2$ is called the residual square sum, y_i is the n-year NPP value, $\hat{y}_i$ is the regression value of the n-year NPP, $\bar{y}$ is the average of the 29-year NPP, and n ($n = 29$) is the number of years.

3.3.4 Correlation Coefficient

In this study, pixel-based spatial analysis was used to analyze the correlation between grassland NPP and climatic factors. The correlation coefficients between annual NPP and annual mean temperature or precipitation were calculated as follows:

$$R_{xy} = \frac{\sum_{i=1}^{n} [(x_i - \bar{x})(y_i - \bar{y})]}{\sqrt{\sum_{i=1}^{n} (x_i - \bar{x})^2 \sum_{i=1}^{n} (y_i - \bar{y})^2}} \tag{3.6}$$

where R_{xy} is the correlation coefficient of the two variables x and y, x_i is the grassland NPP of the i-year, y_i is the annual average temperature or precipitation of the i-year, and $\bar{x}$ is the average of the grassland NPP, $\bar{y}$ is the annual average temperature or precipitation, and i is the sample number ($i = 29$).

3.4 Results and Discussions

3.4.1 Spatial Distribution Characteristics of Grassland NPP in China

From 1982 to 2010, the annual average NPP of China's grassland was 282.0 gC/m²/year, and the spatial distribution generally increased from northwest to southeast across China (Fig. 3.3). Statistical analysis showed that regions with NPP of <50 gC/m²/year accounted for 10.8% of the total grassland area of China; these regions are mainly distributed in Alashan Plateau, southern Xinjiang, northern Tibetan Plateau. Region with average NPP values ranging from 100 to 200 gC/m²/year is the largest and accounted for 21.5% of the case, which mainly distributed in the western Qinghai–Tibet Plateau, the Ordos Plateau, and the central Inner Mongolia. Regions with average NPP values ranging from 200–300 gC/m²/year, 300–400 gC/m²/year, and 400–500 gC/m²/year were dis-

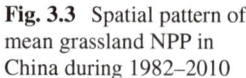

Fig. 3.3 Spatial pattern of mean grassland NPP in China during 1982–2010

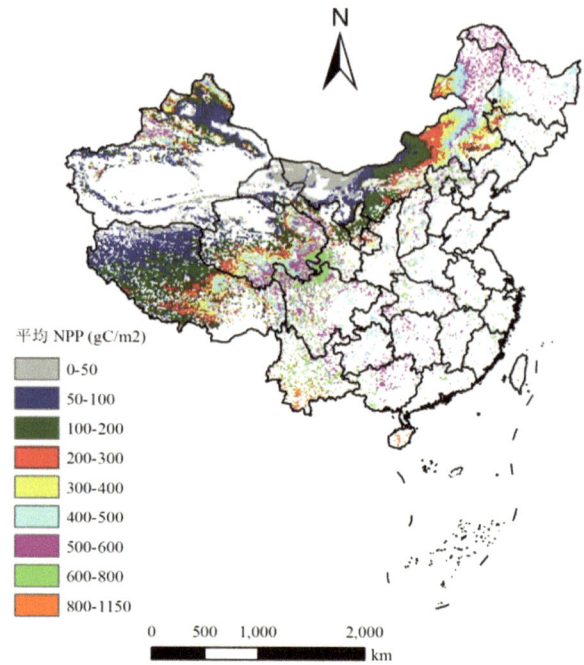

N

平均 NPP (gC/m2)

- 0-50
- 50-100
- 100-200
- 200-300
- 300-400
- 400-500
- 500-600
- 600-800
- 800-1150

0   500   1,000   2,000
                                    km

tributed along the southwest–northeast of China, respectively, with an area of 11.4%, 10.2%, and 10.3% of the total grassland area in China. The grasslands with average NPP of 600–1150 gC/m^2/year were mainly distributed in the southern grass hills and grass slopes, accounting for 9.0% of the total grassland area.

As shown in Table 3.1, the average NPP of 19 grassland types of China was quite different: The highest NPP of dry savanna shrub was 710.2 gC/m^2; the second highest (658.0 gC/m^2) was showed in the hot grass; and the NPP of warm shrub grass was 602.6 gC/m^2, the average NPP of alpine desert (57.7 gC/m^2) was the lowest, and the average NPP of the alpine desert steppe was 63.0 gC/m^2. The average amount of annual grassland NPP in China was 988.3 Tg C (1 Tg $= 10^{12}$ g), of which the total amount of alpine meadow NPP (249.3 Tg C/year) was maximum, accounting for 25.2% of the total; the next was temperate grassland (117.2. Tg C), and the total amount of NPP in mountain meadow was 75.2 Tg C, and the NPP amount of the improved grassland (1 Tg C) was minimum.

3.4.2   Time Dynamic Analysis of Grassland NPP in China

During 1982–2010, the total amount of grassland NPP fluctuated greatly (Fig. 3.4). In 1994, 1998, and 2002, the total amount was 1112.7 Tg C, 1102.7 Tg C, and 1077.3 Tg C, respectively, which was 11.1%, 10.3%, and 8.2% higher than the mul-

Table 3.1 Mean and total NPP for different grassland types

Grassland types	Mean NPP (gC/m²)	Total NPP (Tg C)
Temperate meadow steppe	438.6	71.0
Temperate steppe	292.4	117.2
Temperate desert steppe	161.8	33.3
Alpine meadow steppe	167.1	10.9
Alpine steppe	132.8	59.3
Alpine desert steppe	63.0	7.0
Temperate steppification desert	105.6	11.7
Temperate desert	78.4	37.3
Alpine desert	57.7	3.6
Warm grass	497.5	23.1
Warm shrub grass	549.2	34.2
Hot grass	658	72.5
Hot shrub grass	602.6	76.0
Dry savanna shrub	710.2	4.4
Lowland meadow	337.1	95.0
Mountain meadow	535.4	75.2
Alpine meadow	368.4	249.3
Swamp	396.1	6.4
Improved grassland	391.6	1.0

Fig. 3.4 Inter-annual changing trend of annual total NPP of grassland in China

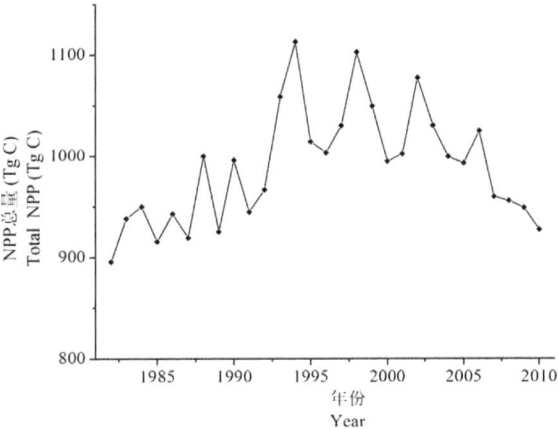

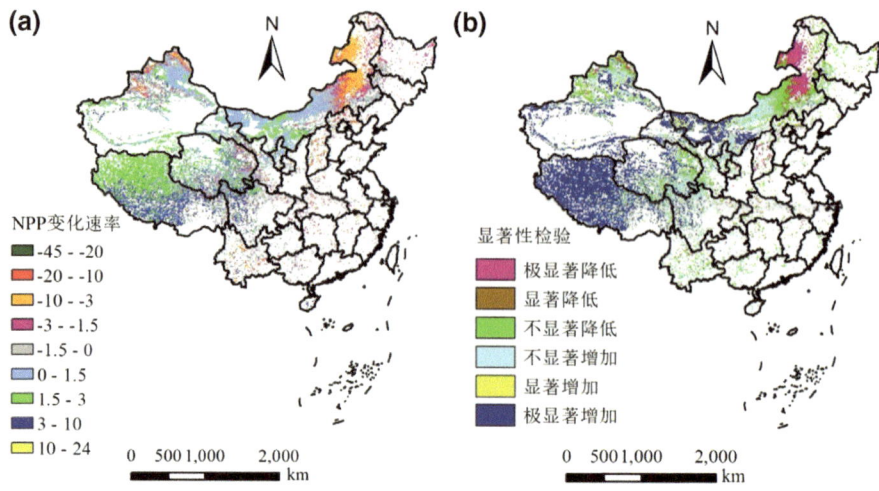

Fig. 3.5 Grassland NPP changing trend (**a**) and significance test (**b**) from 1982 to 2010

tiyear average (988.3 Tg C). In 1982, the total amount of NPP was only 896.0 Tg C, which was the lowest value in 29 years, 10.4% lower than the multiyear average. And the total amount of NPP in 1985 and 1987 was also lower, 915.7 Tg C and 919.3 Tg C, respectively.

From 1982 to 2010, the spatial distribution pattern of the inter-annual variation trend of grassland NPP in China showed that the grassland NPP in China increased with a rate of 0.6 gC/m²/year (Fig. 3.5a). The regions with significant increase of NPP were mainly distributed in the western Qinghai–Tibet Plateau, Alxa Plateau, and Western Xinjiang, while the regions with significant decrease of NPP were mainly distributed in Inner Mongolia of Mongolia Plateau (Fig. 3.5b).

According to the significant statistics of grassland NPP variation (Table 3.2), the area of grassland NPP increased accounted for 67.3% of the total grasslands in China, and the proportion of grassland NPP increased with extremely significance (35.8%) and increased significantly (8.0%) was larger than the ratio which decreased with extremely significance (5.8%) and decreased significantly (4.8%). The significances of NPP variation in 19 grassland categories were significantly different.

3.4.3 Correlation Analysis Between Grassland NPP and Temperature and Precipitation

The variation trend of temperature and precipitation in China from 1982 to 2010 was shown in Fig. 3.6. The annual average temperature and precipitation of Xinjiang, Alxa Plateau, and Qinghai–Tibet Plateau increased, and the trend of warm humidification was obvious. The trend of warm drying was obvious in grass slope distribution

Table 3.2 Statistical analysis of significance test of different grassland type's NPP

Grassland types	Extremely significant decrease	Significant decrease	Insignificant decrease	Insignificant increase	Significant increase	Extremely significant decrease
1	34.1	15	35.9	11.4	1.4	2.2
2	10.3	8.5	38.2	24.7	4.3	14
3	1.7	3.1	23.3	45.8	8.4	17.7
4	0	0.1	3.1	10.1	7.6	79.1
5	0.2	0.3	4.1	10.3	7.4	77.8
6	0	0	0.5	3.6	6	89.8
7	2.1	1.9	10.6	39.6	15.2	30.6
8	0.5	0.8	7.6	24.2	18	49
9	0.1	0.1	1.1	4	5.9	88.8
10	21.3	13	41.7	19.4	2.6	2
11	13.3	12.3	42.8	21.6	4	6
12	4.9	8.8	50.8	29.6	3.1	2.8
13	7.6	11.2	50.6	24.8	3	2.8
14	14.2	8.9	42	29.8	3.6	1.5
15	16.5	11.5	34	15.1	5.3	17.6
16	7.2	8	33.2	32.9	5.8	12.9
17	0.7	1.8	19.5	31.4	8.3	38.3
18	12.9	9.5	37.7	22.6	5.6	11.6
19	14.5	13.3	37.4	28.4	2.5	3.8
20	5.8	4.8	22.2	23.4	8	35.8

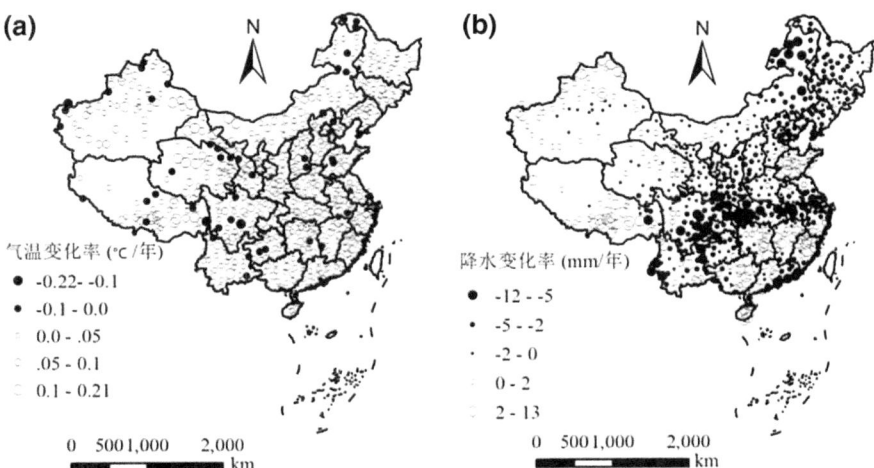

Fig. 3.6 Change trend of annual mean temperature (**a**) and precipitation (**b**) during 1982–2010 in China

area in southern China (Fig. 3.6). In the past 30 years, statistical analysis showed that the temperature in China's grassland distribution area has increased (0.39 °C/10a), and the area with increasing trend accounted for 84% of the total grassland area in China; precipitation has decreased (−3.93 mm/10a), with an area ratio of 56%. Figure 3.7 showed the correlation coefficient between grassland NPP and annual average temperature and precipitation. On the whole, the grassland NPP was positively correlated with annual average temperature ($R = 0.12$). And the correlation coefficient was significantly positive in the southwestern Tibetan Plateau and Alxa Plateau ($P < 0.05$). However, the NPP was negatively correlated with the annual average temperature in Inner Mongolia, the western Tibetan Plateau, the northern Xinjiang, and the grassland–hillside areas of southern China.

The correlation coefficient between grassland NPP and annual precipitation was 0.22. In the spatial distribution, the correlation coefficient was positive in most regions, especially in Inner Mongolia, the Ordos Plateau, and the western Tibetan Plateau, which were significantly positively correlated ($P < 0.05$). However, in the Altai Mountains and the southeastern Tibetan Plateau, the grassland NPP was negatively correlated with precipitation and positively correlated with temperature. The reason may be that the regions belong to cold and humid environment with high altitude, and the low temperature was the limiting factor for vegetation growth. In the past 30 years, the precipitation has increased significantly (Fig. 3.6b); however, the increase in precipitation further reduced the temperature, thereby inhibiting vegetation growth. Therefore, the true mechanism of the negative correlation between grassland NPP and precipitation was affected by temperature.

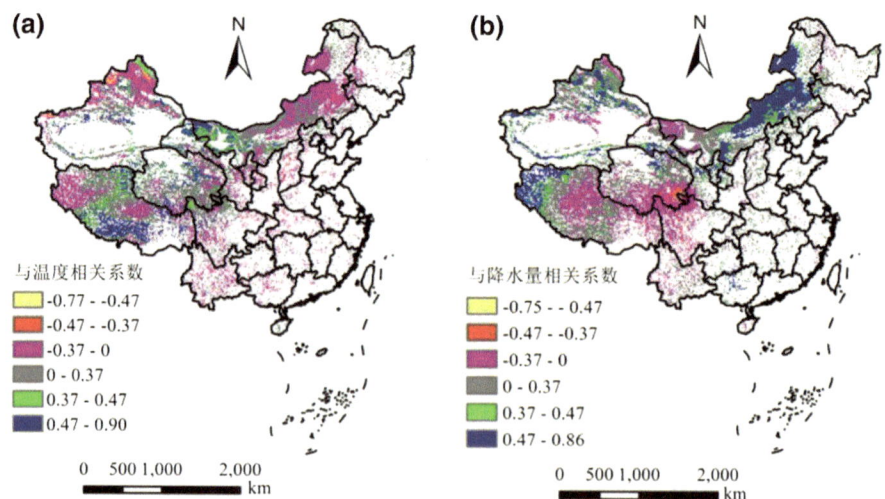

Fig. 3.7 Correlation coefficient between grassland NPP and temperature (**a**) and precipitation (**b**) in China

3.4.4 Discussions

(1) Temporal variation characteristics of grassland NPP in different types

In 1980s and 1990s, the increasing trend of grassland NPP in China decreased gradually, and NPP showed a decreasing trend from 2001 to 2010 (Fig. 3.8). The interdecadal trend differences of NPP variation in 19 grasslands were obvious, and there were four characteristics in general: ① NPP of alpine meadow steppe, alpine desert steppe, alpine desert, alpine steppe, temperate desert, temperate steppification desert, and temperate desert steppe increased from 1982 to 2010. The reasons were as follows: on the one hand, due to the warm and humidification of the climate in the above regions, which was conducive to the growth of grassland vegetation; on the other hand, the implementation of the National Grazing Returning Project, such as artificial grass planting, fence enclosure and rotational grazing measures, promoted the growth of grassland vegetation to a certain extent, resulting in increased productivity. ② The NPP of mountain meadow, alpine meadow, and improved grassland increased from 1982 to 2000 and decreased from 2001 to 2010. ③ The NPP of temperate meadow steppe, temperate steppe, warm grassland, warm shrub grass, lowland meadow, and swamp increased in 1980s, but decreased from 1991 to 2010. The NPP of hot grass, hot shrub grass, dry savanna shrub decreased in the past 30 years. The reason may be that the precipitation in southern China decreased in the past 30 years, temperature increased, and warm dry trend was obvious (Fig. 3.6), which was not conducive to grassland growth and productivity increase.

(2) The impact of hydrothermal factor changes on grassland NPP

Climate change is the main factor affecting the inter-annual variability of terrestrial vegetation activities (Keeling et al. 1996; Weltzin et al. 2003). The results found that: ① from 1982 to 2010, annual mean temperature of 84% regions of grasslands in

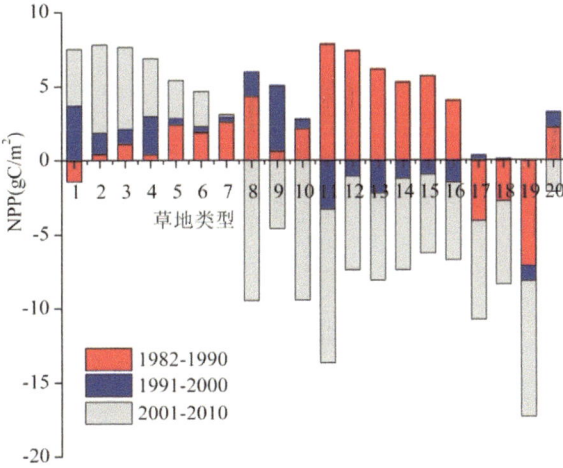

Fig. 3.8 Change trend of annual NPP of different grassland types from 1980s to 2010s. *Note* The meaning of code 1–20 was the same as Fig. 3.1

China showed an increasing trend (0.39 °C/10 year), slightly lower than that in the whole country [0.42 C/10 year]. ② The annual precipitation of 56% grassland distribution areas showed a decreasing trend (−3.9 mm/10 year). ③ The temperature and precipitation in the arid regions of northwestern areas and Tibetan Plateau increased, which was consistent with the previous research on climate change in northwestern China, that is, the climate in northwest China has a warm and humid trend since 1980s (Shi et al. 2007). Because of the increase of precipitation, the stress of water on grassland growth reduced, which promoted the grassland growth in arid areas. ④ The NPP of 67.2% grassland distribution area in China showed an increasing trend, with an increase rate of 0.6 gC/m2/year.

The effect of hydrothermal factors on grassland NPP: The correlation coefficient between NPP and precipitation was greater than that with temperature ($R = 0.22$, 0.12), indicating that the growth of grassland vegetation was more affected by precipitation, because water was the most important limiting factor for grassland life activities and NPP changes, especially in arid regions. As the temperature increased, evapotranspiration increased and available water decreased, making the environment drier (Shen et al. 2012). However, the dependences of grassland NPP on temperature and precipitation were different in different types (Table 3.3). It was found that the correlation coefficient between NPP and precipitation in temperate desert steppe was the largest ($R = 0.44$), which was extremely significant positive correlation. Temperate steppe ($R = 0.4$) and temperate meadow steppe ($R = 0.4$) followed, and reached a significant positive correlation; this was because the above three types grasslands were mainly distributed in the western arid and semi-arid areas, the increase of precipitation in arid regions was conducive to the growth of grassland vegetation (Yang et al. 2008). Yin He et al. (2011) also found that precipitation factors were closely related to vegetation restoration in desert areas in the study of desertification in Inner Mongolia. In the years with high precipitation, the sparse grassland and shrubs on the edge of desert had better growth conditions and vegetation restoration (Yin et al. 2011). The NPP of warm grass, warm shrub grass, hot grass, hot shrub grass, and dry savanna shrub was negatively correlated with temperature. Due to the hot climate in the distribution area of southern grassy slope, the increase of temperature would further lead to the increase of evapotranspiration, which was not conducive to the growth of grassland vegetation.

3.5 Conclusions

(1) The annual average NPP of grassland in China was 282.0 gC/m^2/year, and the total annual NPP was 988.3 Tg C. During the period from 1982 to 2010, grassland NPP showed an increasing trend (0.6 gC/m^2/year). Regions showing increasing NPP accounted for 67.2% of the total grassland areas.

(2) The spatial distribution of grassland NPP showed an increasing trend from the northwest to southeast across China. Clear increases in grassland NPP were observed in the west of the Qinghai–Tibet Plateau, the Alxa Plateau, and western

Table 3.3 Coefficient between grassland NPP and annual mean temperature and precipitation

Grassland types	Coefficient correlation with temperature	Coefficient correlation with precipitation
1	0.03	0.40*
2	0.03	0.40*
3	0.06	0.44**
4	0.25	0.05
5	0.19	0.19
6	0.17	0.36
7	0.12	0.34
8	0.23	0.23
9	0.24	0.37
10	−0.11	0.17
11	−0.06	0.12
12	−0.06	0.15
13	−0.1	0.16
14	−0.13	0.11
15	0.13	0.23
16	0.06	0.03
17	0.2	0.01
18	0.12	0.18
19	0.06	0.27
20	0.12	0.22

Note The meaning of the code of 1–20 was same as Table 3.2; * represent significant correlation ($P < 0.05$), ** represent extremely significant correlation ($P < 0.01$)

area of Xinjiang. Areas with a decrease in grassland NPP were mainly distributed in the western regions of Inner Mongolia.

(3) The increasing rate of NPP showed temporal variation and differed among different grassland types. The grassland NPP increased from 1980s to 1990s and decreased in 2010s. The NPP of alpine grassland and desert grassland increased in the past 30 years, while that of tropical shrub and grassland decreased.

(4) The correlation coefficient between NPP and precipitation was larger than that between NPP and temperature. Moreover, the response of grassland NPP to temperature and precipitation differed for different grassland types. There were significant positive correlations between annual precipitation and NPP in temperate desert steppe, temperate steppe, and temperate meadow steppe ($P < 0.05$). There were negative correlations between annual temperature and NPP in warm grass, warm shrub grass, hot grass, hot shrub grass, and dry savanna shrub.

References

CCICCD (1996). China national committee for the implementation of the UN convention to combat desertification. China country paper to combat desertification. China Forestry Publishing House, Beijing

Christensen L, Coughenour MB, Ellis JE, Chen ZZ (2004) Vulnerability of the Asian typical steppe to grazing and climate change. Clim Change 63(3):351–368

Fang JY, Piao SL, Tang Z, Peng C, Ji W (2001) Interannual variability in net primary production and precipitation. Science 293(5536):1723

Field CB, Randerson JT, Malmström CM (1995) Global net primary production: combining ecology and remote sensing. Remote Sens Environ 51(1):74–88

Keeling CD, Chin J, Whorf TP (1996) Increased activity of northern vegetation inferred from atmospheric CO_2 measurements. Nature 382(6587):146–149

Ma MG, Frank V (2006) Interannual variability of vegetation cover in the Chinese Heihe River Basin and its relation to meteorological parameters. Int J Remote Sens 27(16):3473–3486

Myneni RB, Keeling CD, Tucker CJ, Asrar G, Nemani RR (1997) Increased plant growth in the northern high latitudes from 1981 to 1991. Nature 386(6626):698–702

Ni J (2004) Estimating net primary productivity of grasslands from field biomass measurements in temperate northern China. Plant Ecol 174(2):217–234

Paruelo JM, Epstein HE, Lauenroth WK, Burke IC (1997) ANPP estimates from NDVI for the central grassland region of the United States. Ecology 78(3):953–958

Piao SL, Fang JY, Guo QH (2001) Application of CASA model to the estimation of Chinese terrestrial net primary productivity. Acta Phytoecologica Sinica 25(5):603–608

Piao SL, Fang JY, Ji W, Guo QH, Ke JH, Tao S (2004) Variation in a satellite-based vegetation index in relation to climate in China. J Veg Sci 15(2):219–226

Piao SL, Mohammat A, Fang JY, Cai Q, Feng JM (2006) NDVI-based increase in growth of temperate grasslands and its responses to climate changes in China. Glob Environ Change 16(4):340–348

Piao SL, Fang JY, Zhou LM, Tan K, Tao S (2007) Changes in biomass carbon stocks in China's grasslands between 1982 and 1999. Glob Biogeochem Cycles 21(2):1–10

Potter CS, Randerson JT, Field CB, Matson PA, Vitousek PM, Mooney HA (1993) Terrestrial ecosystem production: a process model based on global satellite and surface data. Glob Biogeochem Cycles 7(4):811–841

Running SW, Nemani RR, Heinsch FA, Zhao M, Reeves M, Hashimoto H (2004) A continuous satellite-derived measure of global terrestrial primary production. Bioscience 54(6):547–560

Scurlock J, Hall DO (1998) The global carbon sink: a grassland perspective. Glob Change Biol 4(2):229–233

Shen WS, Li HD, Sun M, Jiang J (2012) Dynamics of aeolian sandy land in the Yarlung Zangbo River basin of Tibet, China from 1975 to 2008. Glob Planet Change 86:37–44

Shi YF, Shen YP, Kang E, Li DL, Ding YJ, Zhang G (2007) Recent and future climate change in northwest China. Clim Change 80(3–4):379–393

Walker B, Steffen W (1997) IGBP science no. 1: a synthesis of GCTE and related research. IGBP, Stockholm, pp 1–24

Weltzin JF, Loik ME, Schwinning S, Williams DG, Fay PA, Haddad BM (2003) Assessing the response of terrestrial ecosystems to potential changes in precipitation. Bioscience 53(10):941–952

Yang YH, Fang JY, Ma WH, Wang W (2008) Relationship between variability in aboveground net primary production and precipitation in global grasslands. Geophys Res Lett 35(23):L23710

Yu DY, Shi PJ, Han GY, Zhu WQ, Du SQ, Xun B (2011) Forest ecosystem restoration due to a national conservation plan in China. Ecol Eng 37(9):1387–1397

Zhai PM, Zhang XB, Wan H, Pan XH (2005) Trends in total precipitation and frequency of daily precipitation extremes over China. J Clim 18(7):1096–1108

Zhang F, Zhou GS, Wang YH (2008) Dynamics simulaiton of net primary productivity by a satellite data-driven CASA model in innner Mongolia typical steppe, China. J Plant Ecol 32(4):786–797

Zhu WQ, Pan YZ, He H, Yu DY, Hu HB (2006) Simulation of maximum light use efficiency for some typical vegetation types in China. Chin Sci Bull 51(4):457–463

Chapter 4
Inter-annual Variation in Grassland Net Ecosystem Productivity and Its Coupling Relation to Climatic Factors in China

Abstract Grassland carbon (C) sink/source evaluation is important to terrestrial ecosystem C cycling research. In this paper, boreal ecosystem productivity simulator (BEPS), comprising meteorological data, leaf area index, and land-cover-type data, was used to simulate the grassland NEP of China from 1979 to 2008. Regions with NEP of >0 (C sink) accounted for 73.1% of the total grassland area of China. The total C sequestration reached 26.6 TgC yearly, and grassland NEP was positive from 1979 to 2008. The annual changing characteristics were also analyzed. Grassland NEP was positive with carbon sink from June to September, which was negative with carbon source in the remaining months. The carbon conversion efficiency and water-use efficiency of the grassland increased significantly within 30 years. NEP showed positive correlation with precipitation (accounting for 74.2% of the total grassland area was positively correlated) but weakly positive correlation with temperature (50.2% of the case). Furthermore, significant positive correlation was found between grassland NEP and precipitation, especially in northeastern and central Inner Mongolia, and northern Tianshan of Xinjiang, southwestern Tibet, and southern Qinghai Lake. Furthermore, we defined a precipitation differential (PD) parameter to explore the coupling relation between grassland NEP and precipitation. Generally, areas with positive PD are typically a C sinks (72% of the grassland area of IM). However, regions with negative PD are likely a C source (28% of case). Further analysis showed that 69% of regions have positive PD and positive NEP, and 20% of regions have negative PD and negative NEP. This result confirmed that precipitation deficit restrains C sequestration. However, the rest of 11% of regions was sensitive area of carbon sink and carbon source transition. Among 8% of the regions (with negative PD but positive NEP) have insufficient precipitation, although other conditions (i.e., ecological restoration program) are favorable to C sink. Moreover, 3% of the regions (with positive PD but negative NEP) have sufficient precipitation, but have negative NEP because of inappropriate management or low temperature. Thus, appropriate measures that can convert a C source to a C sink are necessary. This paper can serve as a reference for policymakers for the efficient targeted implementation of ecological engineering.

Keywords Net ecosystem productivity · Carbon sink · Coupling relation · Precipitation differential · Water-use efficiency

© Springer Nature Singapore Pte Ltd. 2020   55
W. Zhou et al., *Remote Sensing Monitoring and Evaluation of Degraded Grassland in China*, Springer Geography,
https://doi.org/10.1007/978-981-32-9382-3_4

4.1 Introduction

Over the past 30 years, global warming has increased by 0.2 °C/10 year due to human activity disruption and changes in the natural environment resulting in elevated concentrations of CO_2 and other greenhouse gases (Hansen et al. 2006). Global warming has caused significant changes in terrestrial ecosystems and on the terrestrial carbon (C) cycle (Luo et al. 2007). Grasslands, as one of the most widespread vegetation types, cover approximately 40% of the land surface (Frank et al. 2010). And contain around 30% of global total soil carbon (C) stocks (Anderson 1991). Therefore, grasslands play an important role in the C budget and cycle of the global terrestrial ecosystem (Hunt et al. 2004; Cheng et al. 2010). Grassland ecosystems are mostly distributed in arid and semi-arid climatic zones, where the ecological environment is fragile and sensitive to climate change (Christensen et al. 2004). Meanwhile, the intensity of human disturbance is relatively large. Grassland is a C sink (Diemer and Körner 1998; Oberbauer et al. 1996), and C absorption is more than the emission. However, serious grassland degradation will lead to the loss of grassland C sink function and even become the C source (Li et al. 2006). Therefore, the role of C sequestration function of grassland properly is very important for regional and global C cycle (Geng et al. 2004). The in-depth research of C flux of grassland ecosystem and its response to climatic factors is of great significance for understanding and predicting the C cycle and balance under the background of global changes (Hanson et al. 2000; Raich and Tufekciogul 2000).

C balance is often assessed with net ecosystem productivity (NEP), which is defined as the CO_2 uptake by photosynthesis minus the CO_2 lost by ecosystem respiration (ER) and thus represents the net C accumulation over a given time interval (Randerson et al. 2002). Positive NEP values represent a net C uptake by the ecosystem from the atmosphere, and negative values occur when ecosystems release C to the atmosphere (Zhang et al. 2010). Thus, improved understanding of mechanisms that underlie C dynamics in grassland ecosystems and increase C sink functions is necessary for accurately quantifying global C budgets and mitigating the greenhouse effects. Grassland ecosystems are sensitive to climate change that results in great fluctuations of C budget (Baldocchi et al. 2004; Kim and Verma 1990; Watson 2000; Wever et al. 2002). Grasslands can be a C sink with site-appropriate management (i.e., stocking rates) (Owensby et al. 2006; Zhu 2011) and adequate precipitation. Furthermore, grassland NEP significantly increases with precipitation increase (Rigge et al. 2013). However, precipitation deficits and inter-annual variability often limit NEP; this result causes grasslands to become a net C source during drought years (Meyers 2001; Zhang et al. 2010).

Grasslands account 40% of the national land area in China. Moreover, temperate grassland is the main type. Meanwhile, typical grassland areas in Inner Mongolia (IM) are important and representative of parts of the Eurasian temperate grassland ecosystem (Bai et al. 2008). Located in the typical research field zone of global change research of International Geosphere-Biosphere Program IGBP, IM is one of the region's most sensitive to climate change (Steffen et al. 1992). Thus, IM is the typical region for studying C sink/source and response of C dynamics to climate change. Domestic research has extensively studied grassland C cycle and C source/sink function. However, few works were conducted on grassland NEP at national scale. At present, several models can be used to simulate NEP of the grassland ecosystem, including CENTURY model, BIOME-BGC, TEM, BEPS, LPJ, and VGM models. The remote sensing ecological coupling model (BEPS model) was used to simulate the grassland NEP in this study.

Significant climate changes in temperate grasslands in China over the past 50 years (Qin 2002; Shi 2003) have certainly affected the plant productivity and the C budget of this region (Xiao et al. 1995). Furthermore, the spatial patterns of gross ecosystem productivity are determined by mean annual precipitation (MAP) and mean annual temperature (MAT), whereas NEP is largely explained by MAT (Yu et al. 2013). The temperate grassland NEP of northern China has significant correlation with precipitation during 2001–2010, and C uptake in this region is sensitive to precipitation and drought (Zhang et al. 2014). Other studies found that positive correlations exist among grassland net primary productivity (NPP), NEP, and precipitation (Dai et al. 2016). Furthermore, close relationships exist between grassland productivity and precipitation temporal–spatial dynamics, especially in arid and semi-arid grassland (Bai et al. 2004). However, these studies disregarded the complex connection between NEP and climate factors, including the quantity of precipitation needed to induce a C sink. In this study, we selected IM as the study area to simulate the typical temperate grassland NEP based on remote sensing data and climate data using the boreal ecosystem productivity simulator (BEPS) model. Considering spatial patterns, we then quantitatively analyzed the relationship between NEP and climate factors from 1982 to 2008.

The main objectives of this study are following: (1) to analyze the spatial–temporal dynamics of the grassland NEP in China, (2) to explore the coupling relation between NEP and precipitation and between NEP and temperature at pixel scales, (3) to analyze inter-annual variation in C conversion efficiency (CSE) and rainfall use efficiency (RUE), and (4) to determine the quantity of precipitation needed to induce a C sink. Comprehensively describing the connections between NEP and precipitation and between NEP and temperature may provide insights for improving the suitability of grassland resource and enhancing C sequestration in IM.

4.2 Data Sources

4.2.1 Study Area

Based on the Global Land Cover 2000 dataset (GLC 2003), China's grassland area is 3.35 million km^2, covering approximately 35% of the country's total land area and is mainly distributed in the northwest China and Tibet plateau. About 94% of the total grasslands in China are found in nine provinces, namely IM, Xinjiang, Qinghai, Tibet,

Gansu, Shanxi, Ningxia, Yunnan, and Sichuan. Northwest China is characterized by arid and semi-arid climate and large temperature differences between day and night. The high mountains with high precipitation, such as Altai, Tianshan, Kunlun, and Qilian, block atmosphere circulation and create vast desert basins in the rain shadow, such as Tarim, Junggar, and Qaidam (Shi et al. 2007).

The Inner Mongolia Autonomous Region, which is located between 37.82°–53.82° N and 97.81°–126.80° E with a mean elevation of 1014 m, is ranked as the third largest region in China (Fig. 4.1). IM is characterized by an arid to semi-arid continental climate with strong climatic gradients and varied land-use practices. The region can be divided into three biomes: the arid desert in the west, grassland in the center, and forest in the northeastern region. Grassland is the dominant vegetation type in IM comprising more than 20% of China's total grassland. A strong east-to-west precipitation gradient results in a decrease in annual precipitation from more than 500 mm in eastern IM to less than 100 mm in the western part.

Owing to the large range of precipitation, three major zonal grassland types—meadow steppe, typical steppe, and desert steppe—are distributed along the northeast

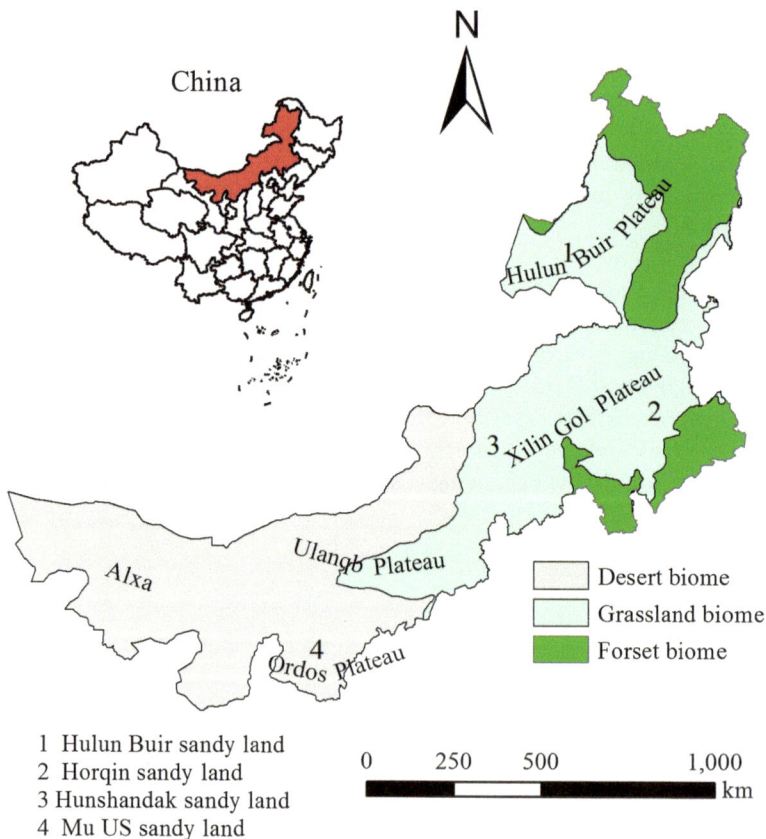

Fig. 4.1 The Inner Mongolia Autonomous Region

to southwest axis in IM. Typical steppe, which developed under semi-arid conditions with annual precipitation from 200 to 400 mm and annual mean temperature from 0 to 8.8 °C in central IM, is the most widely distributed type (Piao et al. 2006). Meadow steppe, which is more productive than a typical steppe, developed in areas with moist fertile soils rich in organic matter in northeastern IM, with annual average precipitation ranging from 300 to 600 mm and annual mean temperature from 2 to 5.8 °C. The desert steppe, which is found in areas with annual precipitation between 150 and 200 mm and annual mean temperature between 5 and 10.8 °C, has the least biomass (John et al. 2009).

4.2.2 Data Source and Processing

Land-use and cover data

Land-cover data used in this study were derived from moderate-resolution imaging spectroradiometer (MODIS) land-cover product. MOD12Q1 data for 2001 with a 1-km resolution were downloaded from United States Geological Survey-Earth Resources Observation and Science (USGS-EROS) (https://lpdaac.usgs.gov/). We used the version with International Geosphere-Biosphere Programme (IGBP) global vegetation classification system. Furthermore, the estimates of accuracy for the dominant IGBP classes in IM were 66% for grassland (Friedl et al. 2010).

LAI data

Leaf area index (LAI) is a key parameter in BEPS for simulating C water flux. In this study, a daily-step 8-km LAI product (CAS, Chinese Academy of Sciences, Global LAI database, Available at: http://www.globalmapping.org/globalLAI/) was used to drive the model. The consistent LAI product was derived from MODIS and advanced very-high-resolution radiometer (AVHRR) products. The MODIS LAI series was generated using GLOBCARBON LAI algorithm for the period since 2000 (2000–2008 in this study). Thereafter, the relationships between AVHRR Ration Vegetation Index (RVI) and MODIS LAI were established pixel by pixel during the overlapped period (2000–2006). On the basis of these relationships, homogeneous LAI series were retrieved from historical AVHRR observations for the period before 2000 (1982–1999 in this study). More details can be found in Liu et al. (2012).

Soil texture data

Soil texture data were collected from the Global Soil Dataset for use in Earth System Models (http://globalchange.bnu.edu.cn/research/soilw). The data are displayed in the volumetric percentages of silt, clay, and sand, from which hydrological parameters, including the wilting point, field capacity, soil porosity, and maximum water-holding capacity, can be estimated. The spatial resolution is 1 km. All the input datasets were extracted and resampled to 8 km to render them fit for model simulation

Climate data

The daily meteorological variables include maximum and minimum air temperature, total precipitation, mean relative humidity, and radiation. We used the global meteorological forcing dataset for land surface modeling (Princeton University, Global Meteorological Forcing Dataset for Land Surface Modeling, available at: http://rda.ucar.edu/datasets/ds314.0/). This database is based on global observation datasets and NCEP/NCAR American National Center for Prediction/ National Center for Atmospheric Research reanalysis, and its accuracy was recently enhanced by incorporating the results from the World Meteorological Organization Solid Precipitation Measurement Intercomparison and Global Precipitation Climatology Project. The data were evaluated using the second Global Soil Wetness Project dataset and 753 basic meteorological stations across China. The spatial resolution is 8 km.

Atmospheric CO_2 data

Monthly atmospheric CO_2 data were collected from Mauna Loa Observatory (MLO), Hawaii (20°N, 156°W; MLO, Air CO_2 database (http://cdiac.esd.ornl.gov/ftp/trends/co2/maunaloa.co2).

4.3 Methodology

4.3.1 BEPS Model Description

The BEPS model is a biogeochemical model (Running and Coughlan 1988) and originally designed to simulate NPP of terrestrial ecosystems at regional or global scales (Liu et al. 1999). The model combines the principles of ecology, biophysics, plant physiology, climatology, and hydrology with the simulation of plant photosynthesis, respiration, C distribution, water, and energy balance. The BEPS model was recently updated to include a module for photosynthesis calculation (Chen et al. 1999) and a CENTURY model-driven soil scheme was incorporated into the model to account for heterotrophic respiration (Rh) and NEP (Ju and Chen 2005; Ju et al. 2006). Although the BEPS model was originally applied to boreal forest ecosystems, the optimization of parameters made it now be applied to forest, grassland, and farmland, etc. ecosystems. In this model, the solar radiation and photosynthesis were calculated by the separation of the sunshine leaf and shade leaf, and the effect of the leaf shape and structure of different vegetation was reduced by the leaf aggregation index. Meanwhile, other biophysical parameters of different vegetation have also been optimized. Recently, it has been used to estimate NPP and NEP across China's landmass through model advancing (Chen et al. 2017; Feng et al. 2007). In this study, BEPS was executed at daily time steps for each pixel, and the annual NEP was obtained as the sum of daily NEP values for China's grassland ecosystem. In this paper, we described only the major methods used in calculating indexes of terrestrial C flux (i.e., NPP, Rh, and NEP) and some parameter modifications. The detailed introduction of the model can be found in Ju et al. (2006). NEP is the difference between net primary productivity (NPP) and R_h, such that

$$NEP = NPP - R_h$$

where NPP is the difference between gross primary productivity (GPP) and autotrophic respiration rate (R_a).

$$NPP = GPP - R_a$$

R_a is simulated as the sum of growth respiration (R_g) and maintenance respiration (R_m). R_g is calculated as a fixed percentage from gross primary productivity (GPP). For R_m simulation, in the prototype model, we replaced the original Bonan algorithm (Bonan 1995) with a plant component-specific scheme.

$$R_{m,k} = T_k \times r_{m,k} \times Q10$$

T_k is the biomass of plant component k, and $r_{m,k}$ is the coefficient of the maintenance respiration. T and T_{ref} are air temperature and reference temperature for maintenance respiration, respectively. $Q10$, r is a temperature sensitivity factor for respiration calculated as a function of temperature (Arora 2003):

$$Q10 = 3.22 - 0.046T$$

R_h is calculated by multiplying the carbon release rate (P_k) from the total carbon amount of soil carbon pool k (C_k).

$$R_h = \sum_{k=1}^{8} P_k C_k$$

According to the grassland ecosystem, the original physiological parameters of the BEPS model were adjusted. The maximum carboxylation rate (V_{max}) is 100 (Li et al. 2011), the specific leaf area index is 25 m^2/kgC, and the respiratory coefficients of leaves, stems, and roots are 0.002, 0.00005, and 0.0002, respectively (Matsushita and Tamura 2002).

4.3.2 Research Indicators and Statistical Analysis

(1) Relationship analysis of NEP and precipitation

In this study, we defined a parameter precipitation differential (PD), which was used to clarify the relation between precipitation and NEP in quantitative and spatial distribution. The calculation of PD was as follows: First, we calculated the basic precipitation (was defined as x-intercept) that represents the amount of precipitation required to achieve an NEP of ≥ 0 gC/m^2/year. The x-intercept of the temporal relationship between mean annual precipitation and average NEP for each pixel during 1979–2008 was calculated using simple least squares linear regression ($n = 27$ at each pixel). Second, this x-intercept value was then subtracted from 1979 to 2008 mean annual precipitation at each pixel, and the resulting product was defined as the PD. Areas with positive PD values indicate that average precipitation is sufficient to maintain a C sink, whereas areas with negative PD would need greater than average precipitation to produce a C sink. The resulting data help to clarify the spatial patterns in the relationship between precipitation and NEP and the likelihood of achieving a positive/negative NEP.

(2) CSE and RUE analysis

In this study, two research indicators selected are C sequestration efficiency (CSE) and RUE.

CSE is the ratio of net primary productivity (NPP) to general primary productivity (GPP), representing the ratio of C accumulated by vegetation to total C assimilation. CSE can also be used as an index to evaluate community health and resistance to diseases and insect pests. When the value is close to 1, the representative plants retain most of the photosynthate products after growth respiration and respiratory maintenance. When CSE is close to 0, the majority of the photosynthates are consumed by living respiration and growth respiration.

RUE is the efficiency of precipitation utilization, which is the ratio of organic matter or NPP to the annual precipitation in the photosynthesis of vegetation. RUE reflects the relationship between plant photosynthesis and water consumption characteristics, which is an extension of the concept of vegetation water-use efficiency. The rate of invalid precipitation is relatively high, and the rate of precipitation variability is large in arid areas. Therefore, RUE is low, and the trend of RUE increases with increasing precipitation.

(3) Statistical analysis

① Simple coefficient correlation analysis

To explore quantitatively the relation between climate factors and grassland NEP, the following formula was used to calculate the simple coefficient correlation.

$$R_{xy} = \frac{\sum_{i=1}^{n}[(x_i - \bar{x})(y_i - \bar{y})]}{\sqrt{\sum_{i=1}^{n}(x_i - \bar{x})^2 \sum_{i=1}^{n}(y_i - \bar{y})^2}}$$

where R_{xy} is the correlation coefficient of variables x and y, x_i is NEP of the ith year, y_i is the temperature or precipitation of the ith year, $\bar{x}$ is the average NEP for all years, $\bar{y}$ is the average temperature or precipitation for all years, and i is the number of years.

② **Partial coefficient correlation analysis**

In order to eliminate the interaction between temperature and precipitation, the partial coefficient correlation between grassland NEP and temperature and precipitation was analyzed. The formula was as follows.

$$R_{123} = \frac{R_{13} - R_{13}R_{23}}{\sqrt{(1 - R_{13}^2) + (1 - R_{23}^2)}}$$

where R_{12} is the correlation between NEP and precipitation, R_{13} is the correlation between NEP and temperature; R_{23} is the correlation between precipitation and temperature.

4.3.3 Implementations and Discussions

4.3.3.1 Accuracy Validation of BEPS Model

C flux tower of temperate grassland in China and field data of C flux originating from previous studies were collected to validate the accuracy of the BEPS model (Sui and Zhou 2013). This paper collected data on domestic flux stations, including carbon flux data from Xilinhot Station and Tongyu Station. We also collected data on grassland carbon flux in the literature, including Xilinhot, Tongyu, Jizhong, and Dongsu research sites. Accuracy validation result is shown in Fig. 4.2. The correlation between field measured NPP and simulated NPP reached significant positive correlation ($R^2 = 0.648$, $P < 0.01$) in Fig. 4.2a. The correlation between the observed NEP and simulated NEP ($R^2 = 0.662$, $P < 0.01$) is illustrated, and the model's estimation accuracy is satisfactory (Fig. 4.2b).

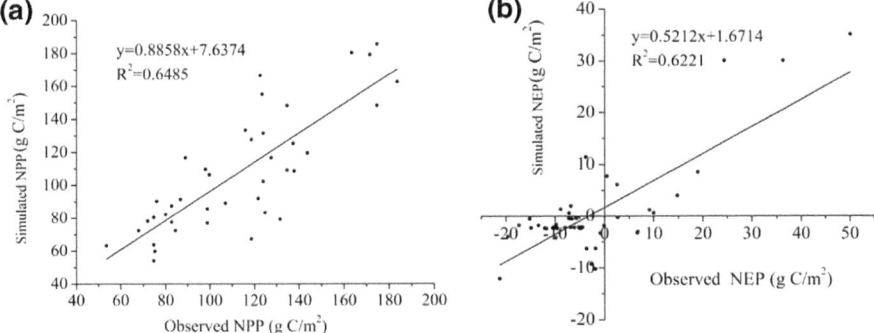

Fig. 4.2 Correlation **a** between field observed grassland net primary productivity (NPP) and simulated, **b** between field observed net ecosystem productivity (NEP) and simulated in Inner Mongolia

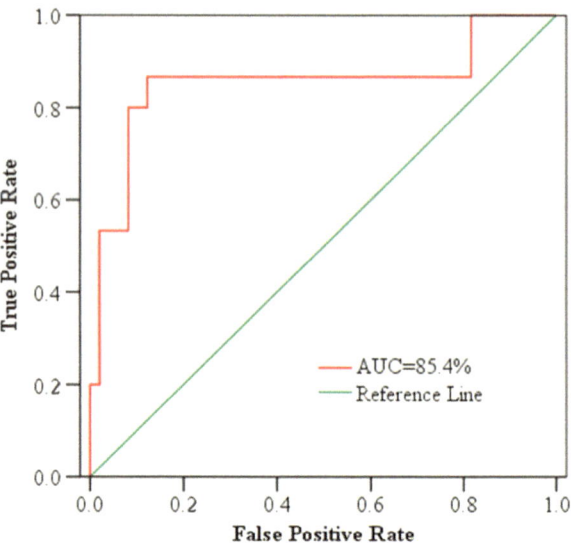

Fig. 4.3 Receiver operator characteristic (ROC) curve and area under the ROC (AUC) curve of grassland carbon sink/source classification confusion matrix

Confusion matrix of carbon sink/source classification

To further verify the confidence of the results obtained from the BEPS model, we validated the C sink/source classification accuracy through a confusion matrix. The receiver operator characteristic (ROC) curve and area under the ROC (AUC) curve validated the model accuracy. In the ROC curve, the vertical axis represents a true positive rate, and the horizontal axis represents a false-positive rate. ROC curve was located in the upper left corner of the graph (Fig. 4.3). Meanwhile, the AUC value was 0.854, which illustrated high classification accuracy.

4.3.4 Spatial Distribution of Grassland NEP in China

Figure 4.4 shows the heterogeneous spatial distribution of grassland NEP in China. In general, grassland NEP distribution increased from northwest to southeast across China, consistent with the distribution of China's isohyet, and the area of the rainfall of >200 mm is NEP > 0. During 1979–2008, the mean annual value of grassland NEP was 13.6 gC/m^2 with weak C sinks.

Statistical analysis showed that regions with NEP of <0 (C source) accounted for 26.9% of the total grassland area of China; these regions are mainly distributed in Hunshandak Sandy Land of Inner Mongolia, the north of Horqin Sandy Land, southern Tibet, northwestern Xinjiang, and the source area of three rivers (Fig. 4.4).

These grasslands are mostly distributed in the arid and semi-arid regions of northwestern China with scarce precipitation and in the puna zone of the Tibetan Plateau with relatively low temperature. In arid regions, the lack of moisture content is not

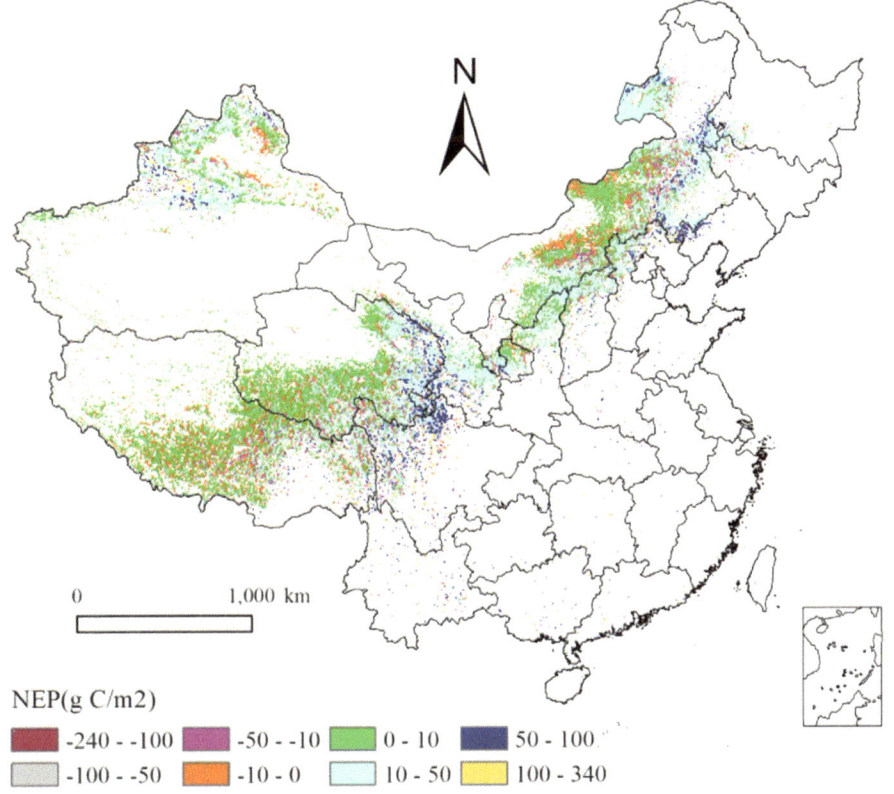

Fig. 4.4 Spatial distribution of grassland net ecosystem productivity (NEP) in China during 1979–2008

conducive to pasture growth and organic matter accumulation. Meanwhile, hyperthermia could lead to enhanced vegetation respiration and accelerated decomposition of organic matter. However, microtherm is unfavorable to the growth of pasture in the northeastern region with adequate precipitation and considered as a limiting factor of NPP accumulation because it affects the vegetation NEP. Microtherm is not conducive to photosynthesis and organic accumulation because of its high elevation in Tibet. Regions with NEP values ranging from -10 to 0 $gC/m^2/year$ accounted for 22.2% of the total grassland area in China. Areas with NEP values ranging from -50–-10 $gC/m^2/year$ accounted for 4.3% of the case and were mainly distributed in the east and south of Hunshandak Sandy Land in the midland of IM, southern Tibet, and western Sichuan. Regions with NEP values of <-50 $gC/m^2/year$ accounted for 0.4% of the total grassland area.

Regions with NEP of >0 (C sink) were widespread in the study area, including central and southern Tibet, Qinghai Province, western Sichuan, southern Gansu, Shaanxi, Shanxi and northern Hebei, Hulun Buir of IM, as well as Tianshan and Altai mountains in Xinjiang. These regions accounted for 73.1% of the total grassland area. Regions with NEP ranging from 0 to 10 gC/m^2/year were located near the 200 mm precipitation line and accounted for 33.5% of the total grassland area; these regions were mainly distributed in central Tibet, the source area of three rivers, and eastern Hunshandak Sandy Land of IM. Regions with NEP ranging from 10 to 50 gC/m^2/year were also widespread in the study area, especially in the eastern part of Qinghai Province, southern Gansu, the northern part of Shaanxi and Shanxi, the Hulun Buir area, and Tianshan area in Xinjiang. These regions accounted for 73.1% of the total grassland area with abundant precipitation. Regions with NEP ranging from 50 to 100 gC/m^2/year accounted for 6.2% of the total grassland area and were distributed in southeastern of Tibetan Plateau, western Sichuan, northern Horqin Sandy Land of IM, and northern slope of Tianshan Mountains in Xinjiang. However, areas with NEP of >100 gC/m^2 were scattered in western Sichuan and accounted for only 1.5% of the total grassland area.

In general, grassland NEP in China was positive, that is, the grassland ecosystem is C sink. However, the grassland NPP decreased owing to the less precipitation and Climate Warming and Drying of the grassland distribution of the northern Hunshandak Sandy Land and Horqin Sandy Land in recent years. However, the respiratory consumption was enhanced by the increase in temperature, leading to decrease in the grassland NEP. The disturbance of human activities, such as overgrazing and grassland reclamation, caused the C sink function of grassland to be reduced and even become C source because of the superposition effect. For the alpine meadow distribution area in southern Tibet, low temperature was considered the limiting factor of grassland growth. Meanwhile, the increase in precipitation enhanced the restriction of low temperatures. The climate change condition was not conducive to organic matter accumulation due to the reduction of grassland NEP.

The annual total grassland NEP in China was 26.6 Tg C, indicating that the annual average net absorption of grassland C in China during 1979–2008 was 26.6 Tg.

4.3.5 Inter-annual and Monthly Changing Trend of Grassland NEP in China

As shown in Fig. 4.4.a, the grassland NEP of China was positive from 1979 to 2008. The amount of CO_2 fixed by grassland photosynthesis is higher than that released by respiration. In the past 30 years, the grassland of China was a C sink in the terrestrial carbon cycle. The trend of inter-annual fluctuation was relatively large, of which the highest NEP of 25.3 gC/m^2 was found in 1988, and the second highest (24.9 gC/m^2) was found in 2003; this finding could be due to the high precipitation in 1988 and the highest precipitation and low temperature in 2003 (Fig. 4.5). In addition, in the implementation of the national program of returning grazing land to grassland in 2003, fencing and grazing measures are conducive to grassland C sink function. The remaining years with high NEPs, such as 1990 and 2005, have low temperatures due to high precipitation (Fig. 4.5). The lowest NEP (4.8 gC/m^2) was found in 1982

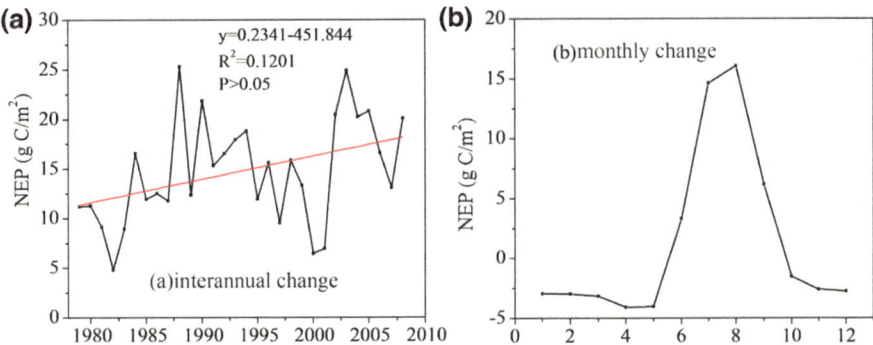

Fig. 4.5 Inter-annual (**a**) and monthly changing trend (**b**) of grassland net ecosystem productivity (NEP) in China

and could be due to the influence of strong El Nino, southern floods, and northern droughts in this year. The low precipitation was not conducive to the photosynthesis and NEP accumulation. The lowest NEP (4.8 gC/m^2) was detected in 2000 owing to the high temperature and low precipitation that increased respiratory consumption and was not conducive to NEP accumulation. Therefore, the comprehensive effect of temperature and precipitation resulted in low NEP.

The variation trend of NEP in every year showed that the NEP in the forage growing season from June to September was positive and showed a C sink (Fig. 4.5). In the rest of months, the grassland NEP from October to May was negative, and the C release was greater than C absorption, showing a C source. The dynamic changes in NEP reflected the seasonal variation in herbage growth. Since the herbage began to enter the grass regreening stage in mid-April, the vegetation NPP was low, but the increasing temperature led to the increase in respiratory consumption. Therefore, the value of NEP was the lowest in April, and the long-time average annual value of NEP in April was -4.1 gC/m^2. Grass began to grow in May, but the organic matter of respiratory consumption was still above the organic matter fixed by photosynthesis, and the average value of NEP was -4.0 gC/m^2. The grassland NEP increased rapidly in June and reached its peak (14.6, 16.1 gC/m^2) in July and August. The NEP began to decline rapidly in September, and the forage entered the withering stage in October with negative NEP (-1.5 gC/m^2).

4.3.6 Inter-annual Variation Trend of Grasslands' CSE and RUE in China

As shown in Fig. 4.6a, the carbon conversion efficiency (CSE) of grassland, that is, the proportion of NPP to GPP, showed a significant increasing trend ($P < 0.001$). This finding indicates that the proportion of vegetation autotrophic respiration decreased and the accumulation of NPP in photosynthates increased. The lowest CSE (0.703)

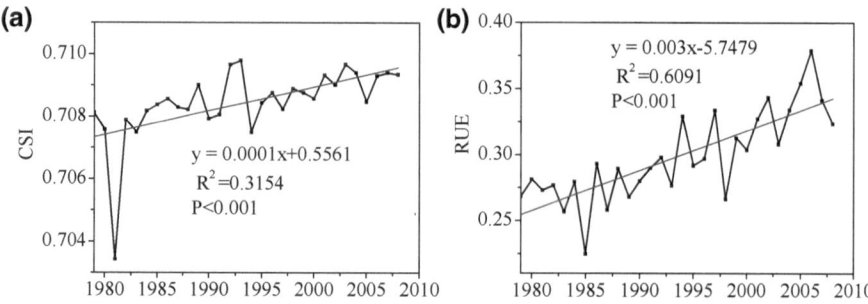

Fig. 4.6 Inter-annual changing tendency of different grasslands' CSE (**a**) and RUE (**b**) in China

was found in 1981, indicating that 70.3% of the photosynthetic products were used for NPP accumulation, and the remaining products were used for the consumption of autotrophic respiration of vegetation. The CSE values of 2004 and 2007 were the highest, indicating the low proportion of organic matter consumed by the self-raised respiration of grassland.

The annual variation of rainwater-use efficiency (RUE) of the grassland showed an extremely significant increase, indicating that the water-use efficiency was significantly improved. As shown in Fig. 4.6b, the average RUE was 0.30, and the peak value reached 0.38 in 2006. The high NPP value increased the ratio between NPP and precipitation. In 1985, the lowest RUE value was 0.23 because low NPP and high precipitation led to low RUE value.

4.3.7 Correlation Analysis of Grassland NEP with Temperature and Precipitation

(1) Simple correlation analysis

Climate factors considerably affect grassland vegetation productivity and C flux. The spatial distribution of correlation coefficient between grassland NEP and precipitation and temperature in China is shown in Fig. 4.7. Regions where NEP has a positive correlation with precipitation were widely distributed (Fig. 4.7a), especially in Inner Mongolia, eastern Qinghai, southwestern Tibet, and northwestern Xinjiang. Statistical analysis showed that the area with positive correlation accounted for 74.2% of the total grassland. The proportion of the regions with significant positive correlation and the positive were 10.1% and 10.5%, respectively (Table 4.1); these regions are mainly distributed in Hulun Buir and the central of Hunshandake Dakenan, southern and eastern Tianshan Mountains, southwestern Tibet, northern Qinghai Lake, and southern regions. Regions with negative correlation between NEP and precipitation were mainly distributed in the three-water headwater region, western Sichuan, and central Tibet. The negative correlation ratio was 25.8%, of which the significant one

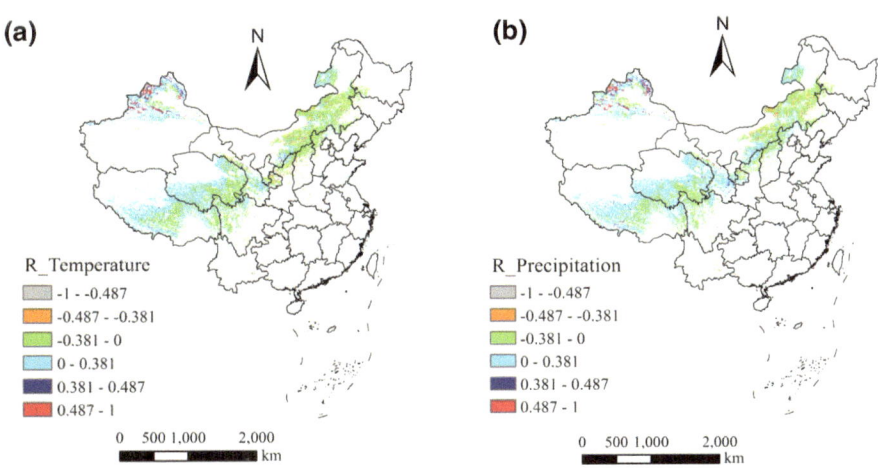

Fig. 4.7 Spatial distribution of simple correlation coefficient between grassland net ecosystem productivity (NEP) and precipitation (**a**) and temperature (**b**)

Table 4.1 Percentage of correlation coefficient significance test between grassland NEP, precipitation, and temperature

Percentage of significant level (%)	Precipitation	Temperature
Extremely significant negative correlation	0.1	1.6
Significant negative correlation	0.8	4.2
Unremarkable negative correlation	24.9	44.1
Unremarkable positive correlation	53.5	42.3
Significant positive correlation	10.5	4.7
Extremely significant positive correlation	10.1	3.2

is only 0.9%. The correlation coefficient of grassland NEP and precipitation is 0.17 in China. The grassland NEP showed a weak positive correlation with temperature, the correlation coefficient was 0.01. The range of NEP was widely distributed and positively correlated with temperature (Fig. 4.7b), and the proportion was 50.2%. The main distributed areas included three-water headwater region, southwest Tibet, northern Shaanxi, and Xinjiang Tianshan Mountains area. Regions with NEP positively related to temperature were mainly distributed in the southwestern Tibet, southwestern of Three-water headwater region, and the proportion was 7.9%. At the same time, regions with NEP negatively correlated with temperature were widespread, accounting for 49.8% of the total grassland area (Table 4.1), mainly distributed in Inner Mongolia, eastern Gansu, northern Shaanxi, eastern Tibet, and northwestern Xinjiang. The precipitation had a significant effect on the increase in grassland NEP. However, the increase in temperature was not conducive to the increase in NEP. This finding could be due to the increase in respiratory consumption caused by the

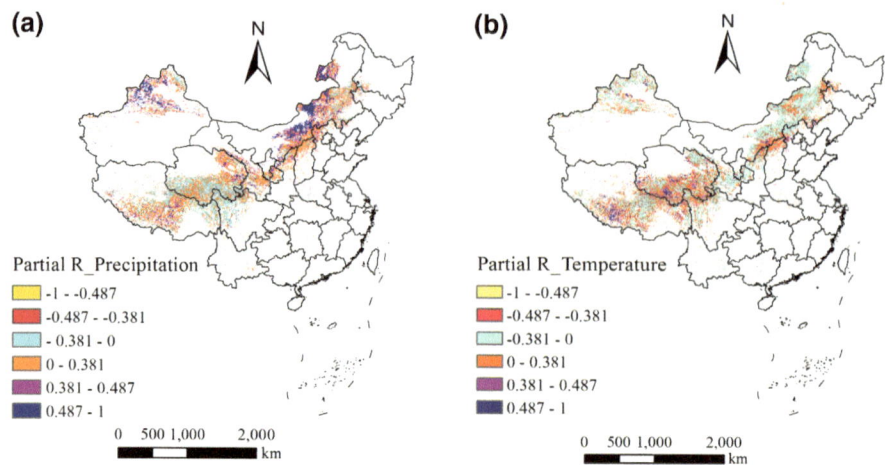

Fig. 4.8 Partial correlation between **a** grassland net ecosystem productivity (NEP) and precipitation and **b** between NEP and temperature

increase in temperature, which was not conducive to the increase in NPP and NEP accumulation.

(2) **Partial correlation analysis**

In order to explore the influence of precipitation and temperature on grassland NEP, separately, the partial correlation coefficient was calculated (Fig. 4.8). Statistical analysis showed that the partial correlation between NEP and precipitation was 0.168, between NEP and temperature was 0.03. And the partial correlation has little difference from the simple correlation coefficient. Positive correlation percentage between NEP and precipitation was 74.8%, and extremely significant positive correlation and significant positive correlation percentage was 9.6% and 10.9%, respectively. Negative correlation percentage reached to 45.8% between NEP and temperature. This result showed that temperature has no effect on the correlation between NEP and precipitation.

4.3.8 Lag Analysis Between Grassland NEP and Precipitation and Temperature

Lag analysis was carried out and results showed in Table 4.2, because previous studies showed that plant growth and biomass accumulation existed obvious lag time from hydrothermal factor (Liang et al. 2014; Zhou et al. 2014). The correlation between grassland NEP and current month' precipitation was negative ($R = -0.003$). But for the positive correlation between NEP and previous months' precipitation, and

Table 4.2 Lag analysis statistic results between NEP and precipitation and temperature

	Current month	One month before	Two months before	Three months before	Four months before
Precipitation	−0.003	0.425	0.707	0.759	0.604
Temperature	−0.033	0.367	0.661	0.787	0.712

the correlation coefficient was largest ($R = 0.759$) between NEP and previous three months' precipitation, the lag time was three months.

The lag time between NEP and temperature was three months ($R = 0.787$). But the correlation between NEP and current month's temperature was negative ($R = -0.033$). Because the soil respiration rate showed a significant positive correlation with soil temperature and air temperature (Cui and Zhang 2016), but experimental warming reduced GPP during the growing seasons (Li et al. 2017). But for the positive correlation between NEP and previous months' temperature, the reason may be that the warming of previous months is beneficial to guarantee the accumulated temperature for the growth of vegetation, and the rise of precipitation guarantees the soil available water content.

4.3.9 Connection Between PD and Grassland NEP

To explore further the connection between precipitation and grassland NEP, we calculated the PD and analyzed the correlation between PD and NEP. The calculation method was provided in the Methods.

As previously mentioned, positive PD values indicate that average precipitation is sufficient to maintain a C sink. Areas with a positive PD (Fig. 4.9a) are typically a C sink and accounted for 72% of the total grassland area. Regions with a negative PD are likely to be a C source under the current precipitation (28% of case). Statistical calculation found that the PD value in IM is 18 mm.

Figure 4.9b shows the spatial pattern between PD and NEP. Regions with positive PD and positive NEP accounted for 69% of the total grasslands. On the contrary, regions with negative PD and negative NEP accounted for 20%. For a negative PD, inducing [by management alone] a C sink in areas with a negative PD and negative NEP would be difficult. Similarly, we can infer that areas with both a positive PD and positive NEP would have relative C stability and would likely require the combination of a drought and heavy grazing to induce a negative NEP. Thus, these regions will be consistent C sink. These findings further prove that grassland NEP is mainly determined by precipitation in IM.

However, 3% of regions have positive PD and negative NEP; the reasons may be that these regions have adequate water conditions but have low temperature or inappropriate management. Additionally, these regions are mainly scattered in the eastern of IM, with high precipitation. Meanwhile, precipitation may not be the

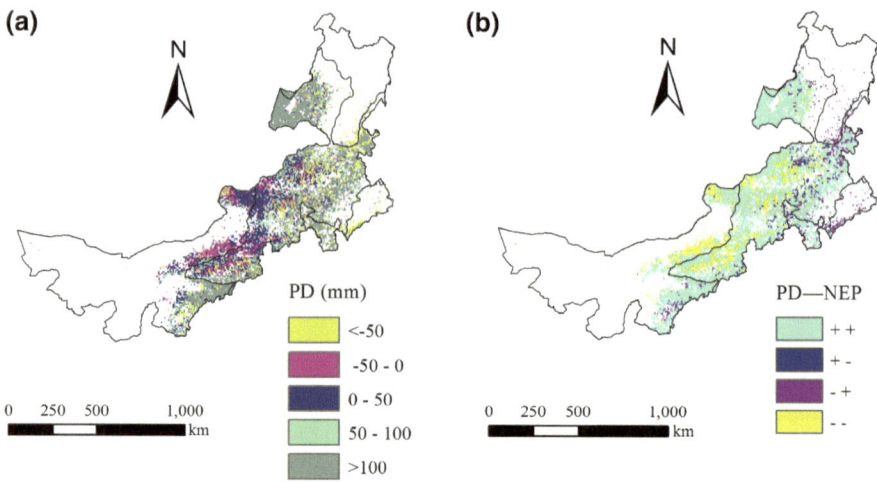

Fig. 4.9 Distribution of **a** precipitation differential (PD) and **b** coupling relation between PD and net ecosystem productivity (NEP) in Inner Mongolia

dominant factor for C sequestration in 8% of regions, which are located in semi-arid areas and have negative PD and positive NEP values. That is to say, grassland NEP is not determined by a single physiological process but rather by a result of the combined responses of climate factors and human management. Therefore, the 11% of regions were defined as sensitive area and fragile zone of carbon source and sink transformation.

4.3.10 Comparisons

(1) **Grassland NEP simulation results comparisons**

For the study of C flux and C cycle in grassland ecosystem, the main research methods include model simulation and site observation. Zhang et al. (2009) used the GLPM model to simulate the average of grassland NEP at 1.91 $gC/m^2/year$ in the extreme arid climate zone of Inner Mongolia. The VIP model was used to simulate the NEP of Leymus chinensis grassland in the Xilin River Basin, Inner Mongolia, and the average value of the NEP was -5.5 $gC/m^2/year$ for 50 years. Sui et al. used the TEM model to simulate NEP in the temperate grassland of China in the past 1951–2007 years, with an average value of 11.25 $gC/m^2/year$, showing a weak C sink. In the past 56 years, the C sink rate increased to 2.2 Tg C/10 yr (Sui and Zhou 2013). Li et al. used the CENTURY model and measurement to study C flux in a fixed plot of the Xilin River Basin. The simulated value of NEP is 19.8 $gC/m^2/year$, and the measured value is 39.7 $gC/m^2/year$ (Li et al. 1998). Li et al. estimated the NEP of alpine shrub and alpine meadow in Qinghai Tibet Plateau by static closed-box gas chromatography in

2004, and the values were 68.31 and 16.75 gC/m^2/year (Li et al. 2010). The annual net carbon exchange rate (NEE) was simulated by eddy covariance in the central grasslands of Mongolia, about −41 gC/m^2/year (Li et al. 2005). Some person used the eddy covariance to estimate NEE of the salinization desert in Fukang, Xinjiang, which was −49 gC/m^2/year, showing a C sink (Liu et al. 2012). Some research estimated the alpine grassland NEP of Tibet Bangor County by measurement in 2000, for 71.12 gC/m^2/year (Zhang et al. 2004). Some research used the BIOME-BGC model to measure the NEP of a degraded grassland in Tongyu, Jilin, for 43.9 gC/m^2/year (Wang et al. 2006). Some research used the flux station data to establish a regression model to simulate the NEP of the meadow in the US Great Plains, for 13.9 gC/m^2/year, showing a C sink (Rigge et al., 2013).

However, due to differences in model selection, research period, study area, and grassland classification criteria, the results of this study are somewhat different from previous studies. Based on the BEPS model, the average value of grassland NEP in China was 15.3 gC/m^2/year in the past 1999–2008 years. The variation range of national NEP was −201 to 263 gC/m^2/year, which shows a weak C sink function. NEP of >0 (C sink) was widespread in China, including southwestern Tibet, Qinghai Province, western Sichuan, Gansu, northern parts of Shaanxi, Shanxi and Hebei, Hulun Buir of IM, as well as Tianshan and Altai mountains in Xinjiang. These regions accounted for 73.1% of the total grassland area (Table 4.3).

Table 4.3 Comparison of simulated net ecosystem productivity (NEP) in this study and that by other researchers

Study area	Research time	Research methods	NEP	Journal title
Xilin River Basin in IM	1980–1989	CENTURY model	19.8	Li et al. (1998)
Xilin River Basin in IM	1980–1989	Measurement method	39.7	Li et al. (1998)
Extreme arid areas in IM	2002	GLPM model	1.91	Zhang et al. (2009)
Xilin River Basin in IM	1958–2007	VIP model	−5.5	Cheng et al. (2010)
Central Mongolia	2003–2004	Vorticity correlation technique	41	Li et al. (2005)
Tongyu station, Jilin	2002–2003	BIOME-BGC	43.9	Wang et al. (2006)
Tibet alpine Grassland	2000	Measurement method	71.12	Zhang et al. (2004)
Shrub of Tibetan Plateau	2004	Measurement method	68.31	Li et al. (2010)
Alpine Meadow of Tibetan Plateau	2004	Measurement method	16.75	Li et al. (2010)
China's temperate grassland	1951–2007	TEM model	11.25	Sui Li et al. (2013)
China's grassland	1999–2008	BEPS model	13.6	This paper

(2) **Effects of climate changes on NEP**

The inter-annual variations in global warming and precipitation have an important impact on the C flux of terrestrial ecosystems, especially for grassland ecosystems in arid and semi-arid areas. In arid and semi-arid regions, the change in precipitation is more significant than temperature increasing and CO_2 concentration change on the C flux of the ecosystem (Potts et al. 2006; Weltzin et al. 2003), owing to the growth of vegetation and the productivity of net ecosystem determined by the availability of water (Fang et al. 2001; Knapp et al. 2002; Weltzin et al. 2003). The influence of inter-annual variation of precipitation and temperature on the C flux and C balance of grassland has also been extensively studied worldwide.

In this paper, the BEPS model was used to simulate the grassland NEP in China, and the response of NEP to the change in temperature and precipitation was analyzed. The inter-annual variation trend of NEP and precipitation was consistent, and their values were both higher in 2003. However, no noticeable difference was found between the trend of NEP and temperature change, that is, increasing the temperature is not conducive to C absorption and NEP rose. In this study, the regression analysis of grassland NEP and precipitation showed that, when precipitation in China's grassland distribution area was 292 mm, NEP was 0. Therefore, when precipitation exceeds the value, the probability of NEP being positive is high. Statistical analysis showed that NEP was positively correlated with precipitation ($R = 0.17$), the correlation obviously existed spatial heterogeneity. Significant positive correlation was found in the Hulun Buir of northeastern IM, the southern and eastern regions of Hunshandake in central IM, the north slope of Tianshan Mountains in Xinjiang, southwestern Tibet, and the south areas of Qinghai Lake; this conclusion is consistent with those reported at home and abroad. Research about grassland NEP of the US Great Plains responded to precipitation change found that grassland NEP rose significantly with precipitation increasing, and the correlation between NEP and precipitation in arid area was more visible. Niu et al. conducted a control experiment by controlling climate change and analyzed the response of grassland C flux to changes in temperature and precipitation in Duolun County of IM. The results showed that the increased precipitation would promote the increase in C and water fluxes but reduce the temperature increase and adversely affect the grassland NEP (Niu et al. 2008). However, for high altitude and low temperature of the Tibetan Plateau, precipitation increasing further exacerbated the limitation of low temperature on productivity and is also not conducive to the increase in NEP. As shown in Fig. 4.10, the size of GPP and NPP had the highest degree of explanation for the change in NEP, that is, R^2 was the highest, having reached 0.8 and 0.79. The interpretation degree of precipitation to NEP change was small, R^2 was 0.2, which of temperature was only 0.06. Relation between NEP and temperature showed a weak correlation ($R = 0.01$), and areas that temperature and grassland NEP that were negatively correlated accounted for 49.8% of grassland distribution, due to temperature increasing would promote evapotranspiration and vegetation transpiration, resulting in a decrease in soil water content, which is not conducive to C accumulation (Harte et al. 1995; Wan et al. 2002) and is consistent with Jiang's conclusions for the study of meadow grassland in Song-

Fig. 4.10 Explanation (R^2) of changes in net primary productivity (NPP), gross primary productivity (GPP), precipitation and temperature on the inter-annual change in net ecosystem productivity (NEP)

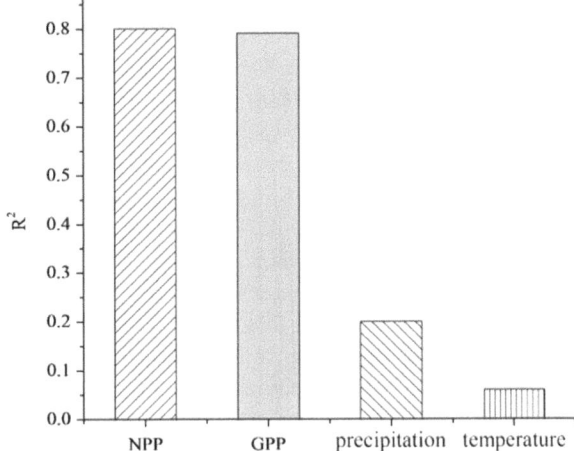

nen Plain. The increase in temperature led to the decrease in NEE and ecosystem respiration. The soil water content is the primary driving factor of grassland C flux, especially in dry years (Jiang et al. 2012). Hao et al. also found that drought stress reduced C accumulation in IM grassland, making the grassland ecosystem change from C sink to C source.

(3) Carbon source/sink sensitive area

In this study, precipitation was the dominant factor that maintained C sinks in IM. The reason may be that NEP is positively correlated with precipitation ($R = 0.31$, 90% of total grassland area) but negatively correlated with temperature ($R = -0.11$, 72% of case). This result is consistent with previous studies in IM (Dai et al. 2016). Furthermore, NEP significantly increased with precipitation (Rigge et al. 2013). Meanwhile, in the past three decades, precipitation significantly decreased ($P < 0.05$), but temperature increased significantly (0.64 °C/10y, $P < 0.001$). In the Songnen grassland in northeast China, warming decreased NEP and increased ER (Jiang et al. 2012).

To understand the interactions between precipitation and NEP, we calculated the PD and evaluated the relationship between PD and NEP. Results showed that 69% of regions have both positive PD and positive NEP, thereby confirming that the regions consistently remain as C sink until the balance is disturbed, particularly by long-term drought or inappropriate management (overgrazing and return grass to farmland). Therefore, these regions are attracting interest from various fields, particularly biofuel production and C sequestration opportunities (Li et al. 2005). However, precipitation deficit restrains C sequestration in these regions (both with negative PD and negative NEP, 20% of case). In central Mongolia, water stress was observed in late July and early August and switched the steppe from a C sink to a C source (Li et al. 2005). In these regions, switching a C source to a C sink is difficult under the current warm–dry climate situation, unless appropriate human management is applied for a long period.

However, 11% of the regions have contradicting PD and NEP values. The regions (with positive PD but negative NEP, 3% of total grassland area) have sufficient precipitation for C sequestration. The negative NEP may be attributed to low temperatures or inappropriate management. This finding needs appropriate measures to return a C source to a C sink and increase C sequestration. By contrast, for regions (with negative PD but positive NEP, 8% of case) with inappropriate precipitation, other combined conditions, such as suitable temperature and appropriate human activities, may be favorable for C sink. Thus, sustainable protection and appropriate management are necessary for maintaining the current C sink. Since 2001, the ecological restoration programs, such as RGGP, have been implemented in China.

Social statistics shows that the rural population has increased slowly since 1986 and slightly decreased since 1995. However, total population continues to grow rapidly (Fig. 4.11). Population urbanization decreases human disturbance to grassland ecosystem. Additionally, the physical and chemical properties of the soil were

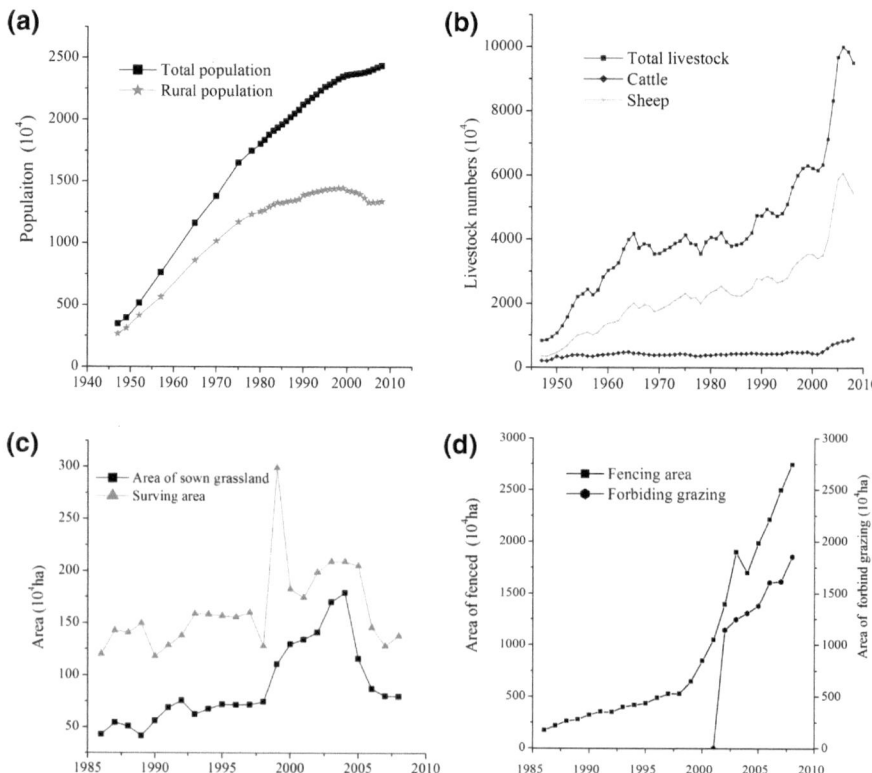

Fig. 4.11 Total and rural population (1947–2009) (**a**), livestock numbers (1947–2009) (**b**), area of sown grassland and surviving area of planted grassland (1986–2009) (**c**), area of fenced grassland and forbidding grazing area (1986–2009) (**d**). *Data sources* Inner Mongolia statistical yearbook (1986–2010)

also improved when human disturbance decreased, thereby increasing soil C sequestration potential and decreasing soil respiration intensity (Cao and Wang 2010). Social statistics also shows that banning grazing and fencing areas have increased the area of sown grassland since 1986. Thus, this ban promotes the increase of C sequestration (Wang et al. 2011). In this study, banning of grazing area reached up to 150 million hectares, and the artificial grasslands area will have reached 30 million hectares by 2020 by increasing C storage by 0.24 Pg C/year.

Meanwhile, in previous studies, the RGGP promoted the increase of grassland coverage and productivity (Zhou et al. 2013), and appropriate grazing density is beneficial for C sequestration. Thus, efforts are necessary for maintaining C sinks function.

4.4 Conclusions

(1) In 1979–2008, the average NEP value of grassland in China is 13.6 gC/m^2, showing a weak C sink. Regions with NEP of >0 (C sink) accounted for 73.1% of the total grassland area of the total grassland area, and the annual net carbon absorption of grassland is 26.6 Tg C. The grassland NEP of in China was positive from 1979 to 2008. The trend of inter-annual fluctuation was relatively large, of which the highest NEP in 1988 was 25.3 gC/m^2, the second average was 24.9 gC/m^2 in 2003, which corresponded to the higher precipitation. The annual variation showed that grassland NEP was positive from June to September, showing C sequestration, while the NEP in the other months was negative, showing C source.

(2) The C conversion efficiency (NPP/GPP) of grassland increased significantly in 1979–2008. Self-consumption of grassland decreased gradually, and the water-use efficiency increased significantly. The grassland NEP showed a positive correlation with precipitation and a weak correlation with temperature. The rates were 74.2% and 50.2%, respectively. The increase in precipitation had noticeable effects on grassland NEP, especially in the northeastern and central IM, northwestern Xinjiang, southwestern Tibet, and Qinghai Lake Region, grassland NEP was positive correlated with precipitation significantly.

(3) We then defined a parameter PD to explore the coupling relation between grassland NEP and precipitation. Areas with a positive PD are typically a C sink and accounted for 72% of the total grassland area. Areas with a negative PD were likely a C source under the current precipitation (28% of case). Statistical results show that the PD value in IM is 18 mm greater than zero. Therefore, the PD value shows that the IM is C sink.

In IM, 69% of the regions have both positive PD and positive NEP, indicating that the regions consistently remain a C sink until the balance is disturbed, especially by long-term drought or inappropriate management. In regions with negative PD and negative NEP (20% of case), precipitation deficit restrains C sequestration. Meanwhile, 11% of the regions have contradicting PD and NEP values, and these regions were defined as sensitive area of carbon source and sink transition. Regions with negative PD but positive NEP account for 8% of the study area and have insufficient precipitation. By contrast, other conditions (human activities) are favorable for C sink. One such condition is the ecological restoration program (RGGP) promoted grassland vegetation growth. Meanwhile, the region (with positive PD but negative NEP) accounts for 3% of the total grassland area, where precipitation is sufficient. The negative NEP values may be attributed to low temperatures or inappropriate management. Thus, these regions need appropriate measures to return a C source to a C sink.

(4) This paper can contribute to an in-depth understanding of C sequestration and provide policy makers some references for the efficient targeted implementation of ecological engineering, such as the sensitive region of carbon sink and carbon source transition drawing.

References

Anderson JM (1991) The effects of climate change on decomposition processes in grassland and coniferous forests. Ecol Appl 1(3):326

Arora VK (2003) Simulating energy and carbon fluxes over winter wheat using coupled land surface and terrestrial ecosystem models. Agric For Meteorol 118(1):21–47

Bai YF, Han XG, Wu JG, Chen ZZ, Li LH (2004) Ecosystem stability and compensatory effects in the Inner Mongolia grassland. Nature 431(7005):181–184

Bai Y, Wu J, Xing Q, Pan Q, Huang J, Yang D (2008) Primary production and rain use efficiency across a precipitation gradient on the Mongolia plateau. Ecology 89(8):2140–2153

Baldocchi DD, Xu L, Kiang N (2004) How plant functional-type, weather, seasonal drought, and soil physical properties alter water and energy fluxes of an oak—grass savanna and an annual grassland. Agric For Meteorol 123(1–2):13–39

Bonan GB (1995) Land-atmosphere CO_2 exchange simulated by a land surface process model coupled to an atmospheric general circulation model. J Geophys Res Atmos 100(D2):2817–2831

Cao SX, Wang GS (2010) Damage caused to the environment by reforestation policies in arid and semi-arid areas of China. Ambio 39(4):279–283

Chen JM, Liu J, Cihlar J, Goulden ML (1999) Daily canopy photosynthesis model through temporal and spatial scaling for remote sensing applications. Ecol Model 124(2):99–119

Chen YZ, Ju WM, Groisman P, Li JL, Propastin P, Xu X (2017) Quantitative assessment of carbon sequestration reduction induced by disturbances in temperate Eurasian steppe. Environ Res Lett 12(11):115005

Cheng Q, Mo X, Wang Y, Lin Z (2010) Simulation of the carbon cycle in the meadow steppe dominated by Leymus Chinensis. Nat Resour J 25(1):60–70

Christensen L, Coughenour MB, Ellis JE, Chen ZZ (2004) Vulnerability of the Asian typical steppe to grazing and climate change. Clim Change 63(3):351–368

Cui H, Zhang YH (2016) Diurnal and seasonal dynamic variation of soil respiration and its influencing factors of different fenced enclosure years in Desert Steppe. Environ Sci 37(4):1507–1515

Dai E, Huang Y, Wu Z, Zhao D (2016) Spatial-temporal features of carbon source-sink and its relationship with climate factors in Inner Mongolia grassland ecosystem. Acta Geogr Sin 71(1):21–34

Diemer M, Körner C (1998) Transient enhancement of carbon uptake in an alpine grassland ecosystem under elevated CO_2. Arct Alp Res 30(4):381–387

Fang JY, Piao SL, Tang Z, Peng C, Ji W (2001) Interannual variability in net primary production and precipitation. Science 293(5536):17–23

Feng X, Liu G, Chen JM, Chen M, Liu J, Ju WM (2007) Net primary productivity of China's terrestrial ecosystems from a process model driven by remote sensing. J Environ Manage 85(3):563–573

Frank AB, Sims PL, Bradford JA, Mielnick PC, Dugas WA, Mayeux HS (2010) Carbon dioxide fluxes over three Great Plains grasslands. In: Follett RF, Kimble JM, Lal R (eds) The potential of U.S. grazing lands to sequester carbon and mitigate the greenhouse effect. Lewis, New York, pp 167–188

Friedl MA, Sulla-Menashe D, Tan B, Schneider A, Ramankutty N, Sibley A (2010) MODIS Collection 5 global land cover: algorithm refinements and characterization of new datasets. Remote Sens Environ 114(1):168–182

Geng Y, Dong Y, Qi Y (2004) Review about the carbon cycle researches in grassland ecosystem. Prog Geogr 23(3):74–81

GLC (2003) Global landcover classification for the year 2000. http://www-gem.jrc.it/glc2000/

Hansen J, Sato M, Ruedy R, Lo K, Lea DW, Medinaelizade M (2006) Global temperature change. Proc Natl Acad Sci USA 103(39):14288–14293

Hanson PJ, Edwards NT, Garten CT, Andrews JA (2000) Separating root and soil microbial contributions to soil respiration: a review of methods and observations. Biogeochemistry 48(1):115–146

Harte J, Torn MS, Chang FR, Feifarek B, Kinzig AP, Shaw R (1995) Global warming and soil microclimate: results from a meadow-warming experiment. Ecol Appl 5(1):132–150

Hunt JE, Kelliher FM, Mcseveny TM, Ross DJ, Whitehead D (2004) Long-term carbon exchange in a sparse, seasonally dry tussock grassland. Glob Change Biol 10(10):1785–1800

Jiang L, Guo R, Zhu T, Niu X, Guo J, Sun W (2012) Water- and plant-mediated responses of ecosystem carbon fluxes to warming and nitrogen addition on the songnen grassland in Northeast China. PLoS ONE 7(9):e45205

John R, Chen J, Lu N, Wilske B (2009) Land cover/land use change in semi-arid Inner Mongolia: 1992–2004. Environ Res Lett 4(4):45010

Ju W, Chen JM (2005) Distribution of soil carbon stocks in Canada's forests and wetlands simulated based on drainage class, topography and remotely sensed vegetation parameters. Hydrol Process 19(1):77–94

Ju W, Chen JM, Black TA, Barr AG, Liu J, Chen B (2006) Modelling multi-year coupled carbon and water fluxes in a boreal aspen forest. Agric For Meteorol 140(1–4):136–151

Kim J, Verma SB (1990) Carbon dioxide exchange in a temperate grassland ecosystem. Bound-Layer Meteorol 52(1):135–149

Knapp AK, Fay PA, Blair JM, Collins SL, Smith MD, Carlisle JD (2002) Rainfall variability, carbon cycling, and plant species diversity in a mesic grassland. Science 298(5601):2202–2205

Li L, Liu X, Chen Z (1998) Study on the carbon cycle of Leymus Chinensis stppe in the Xilin River Basin. Chin J Plant Ecol 40(10):955–961

Li SG, Asanuma J, Eugster W, Kotani A, Liu JJ, Urano T (2005) Net ecosystem carbon dioxide exchange over grazed steppe in central Mongolia. Glob Change Biol 11(11):1941–1955

Li CZ, Ma MG, Zhang F, Jiang ZR (2006) The dynamic analysis of vegetation pattern in the Northwest of China. Remote Sens Technol Appl 21(4):332–337

Li D, Cao G, Huang Y, Jin D, Ming Z (2010) Carbon budget of alpine shrub meadow ecosystem in Qinghai-Tibetan plateau. Acta Prataculturae Sinica 27(1):37–41

Li L, Vuichard N, Viovy N, Ciais P, Ceschia E, Jans W (2011) Importance of crop varieties and management practices: evaluation of a process-based model for simulating CO_2 and H_2O fluxes at five European maize (Zea mays L.) sites. Biogeosci Discuss 8(2):2913–2955

Li GY, Han HY, Du Y, Hui DF, Xia JY, Niu SL (2017) Effects of warming and increased precipitation on net ecosystem productivity: a long-term manipulative experiment in a semiarid grassland. Agric For Meteorol 232:359–366

Liang Y, Ganzhu Z, Zhang WN, Gao QZ, Danjiu L, Xirao Z (2014) A review on effect of climate change on grassland ecosystem in China. J Agric Sci Technol 16(2):1–8

Liu J, Chen JM, Cihlar J, Chen W (1999) Net primary productivity distribution in the BOREAS region from a process model using satellite and surface data. J Geophys Res Atmos 104(27):727–735

Liu R, Li Y, Wang QX (2012) Variations in water and CO_2 fluxes over a saline desert in western China. Hydrol Process 26(4):513–522

Luo Y, Sherry R, Zhou X, Wan S (2007) Terrestrial carbon-cycle feedback to climate warming: experimental evidence on plant regulation and impacts of biofuel feedstock harvest. Annu Rev Ecol Evol Syst 38(38):683–712

Matsushita B, Tamura M (2002) Integrating remotely sensed data with an ecosystem model to estimate net primary productivity in East Asia. Remote Sens Environ 81(1):58–66

Meyers TP (2001) A comparison of summertime water and CO_2 fluxes over rangeland for well watered and drought conditions. Agric For Meteorol 106(3):205–214

Niu S, Wu M, Han Y, Xia J, Li L, Wan S (2008) Water-mediated responses of ecosystem carbon fluxes to climatic change in a temperate steppe. New Phytol 177(1):209–219

Oberbauer SF, Gillespie CT, Cheng W, Sala A, Gebauer R, Tenhunen JD (1996) Diurnal and seasonal patterns of ecosystem CO_2 efflux from upland tundra in the foothills of the Brooks Range, Alaska, USA. Arctic Alp Res, 328–338

Owensby CE, Ham JM, Auen LM (2006) Fluxes of CO_2 from grazed and ungrazed tallgrass prairie. Rangeland Ecol Manag 59(2):111–127

Piao SL, Mohammat A, Fang J, Cai Q, Feng J (2006) NDVI-based increase in growth of temperate grasslands and its responses to climate changes in China. Glob Environ Change 16(4):340–348

Potts D, Huxman TB, Weltzin J, Williams D (2006) Resilience and resistance of ecosystem functional response to a precipitation pulse in a semi-arid grassland. J Ecol 94(1):23–30

Qin DH (2002) Assessment on environment of Western China (Synopsis). Science Press, Beijing

Raich JW, Tufekciogul A (2000) Vegetation and soil respiration: correlations and controls. Biogeochemistry 48(1):71–90

Randerson JT, Iii FSC, Harden JW, Neff JC, Harmon ME (2002) Net ecosystem production: a comprehensive measure of net carbon accumulation by ecosystems. Ecol Appl 12(4):937–947

Rigge M, Wylie B, Zhang L, Boyte SP (2013) Influence of management and precipitation on carbon fluxes in great plains grasslands. Ecol Ind 34:590–599

Running SW, Coughlan JC (1988) A general model of forest ecosystem processes for regional applications I. Hydrologic balance, canopy gas exchange and primary production processes. Ecol Modell 42(2):125–154

Shi Y (2003) Discussion on the present climate change from warm dry to warm wet in northwest China. Quat Sci 23(2):152–164

Shi YF, Shen YP, Kang E, Li DL, Ding YJ, Zhang G (2007) Recent and future climate change in northwest China. Clim Change 80(3):379–393

Steffen WL, Walker BH, Ingram JSL (1992) Global change and terrestrial ecosystems: the operational plan. Paper presented at the Global Change Report, No. 21, Sweden

Sui XH, Zhou GS (2013) Carbon dynamics of temperate grassland ecosystems in China from 1951 to 2007: an analysis with a process-based biogeochemistry model. Environ Earth Sci 68(2):521–533

Wan S, Luo Y, Wallace LL (2002) Changes in microclimate induced by experimental warming and clipping in tallgrass prairie. Glob Change Biol 8(8):754–768

Wang XM, Chen FH, Dong ZB (2006) The relative role of climatic and human factors in desertification in semiarid China. Glob Environ Change 16(1):48–57

Wang X, Zang S, Na X (2011) Analyzing dynamic process of land use change in Ha-Da-Qi industrial corridor of China. Procedia Environ Sci 11(Part B): 1008–1015

Watson RT (2000) Land use, land-use change, and forestry: a special report of the intergovernmental panel on climate change. Cambridge University Press, Cambridge

Weltzin JF, Loik ME, Schwinning S, Williams DG, Fay PA, Haddad BM (2003) Assessing the response of terrestrial ecosystems to potential changes in precipitation. Bioscience 53(10):941–952

Wever LA, Flanagan LB, Carlson PJ (2002) Seasonal and interannual variation in evapotranspiration, energy balance and surface conductance in a northern temperate grassland. Agric For Meteorol 112(1):31–49

Xiao X, Ojima DS, Parton WJ, Chen Z, Chen D (1995) Sensitivity of Inner Mongolia Grasslands to climate change. J Biogeogr 22(4/5):643–648

Yu GR, Zhu XJ, Fu YL, He HL, Wang QF, Wen XF (2013) Spatial patterns and climate drivers of carbon fluxes in terrestrial ecosystems of China. Glob Change Biol 19(3):798

Zhang X, Shi P, Liu Y, Ouyang H (2004) CO_2 emission and carbon balance of soil in Alpine steppe ecosystem in Tibetan Plateau. Sci China 34(S2):193–199

Zhang N, Zhao YS, Yu GR (2009) Simulated annual carbon fluxes of grassland ecosystems in extremely arid conditions. Ecol Res 24(1):185–206

Zhang L, Wylie BK, Ji L, Gilmanov TG, Tieszen LL (2010) Climate-driven interannual variability in net ecosystem exchange in the northern Great Plains grasslands. Rangeland Ecol Manag 63(1):40–50

Zhang L, Guo H, Jia G, Wylie B, Gilmanov T, Howard D (2014) Net ecosystem productivity of temperate grasslands in northern China: an upscaling study. Agric For Meteorol 184(1):71–81

Zhou W, Li JL, Mu SJ, Gang C, Sun ZG (2013) Effects of ecological restoration-induced land-use change and improved management on grassland net primary productivity in the Shiyanghe River Basin, north-west China. Grass Forage Sci 69:596–610

Zhou W, Gang CC, Chen YZ, Mu SJ, Sun ZG, Li JL (2014) Grassland coverage inter-annual variation and its coupling relation with hydrothermal factors in China during 1982–2010. J Geog Sci 24(4):593–611

Zhu Z (2011) Baseline and projected future carbon storage and greenhouse-gas fluxes in the Great Plains region of the United States. U.S. Geological Survey professional paper; no. 1787

Chapter 5
The Variation of Landscape and NPP of Main Pastoral Grasslands in China

Abstract To explore the grassland ecosystem productivity and landscape ecological patterns of main pastoral grasslands in China, it provides a theoretical basis for the efficient implementation of ecological engineering and rational management of grassland resources in the region. This study analyzed the changes in grassland area, landscape index (LSI), and net primary productivity (NPP) in seven major pastoral areas in China in 1985, 1995, 2005, and 2015. Results showed that (1) the sizes of the grassland study area in 1985, 1995, 2005, and 2015 were 248.34, 243.93, 245.80, and 244.660 km^2, respectively. (2) The dominance of grassland in the landscape pattern increased from 2005 to 2015 as compared with that in 1985–1995 and 1995–2005. The degrees of spatial heterogeneity were reduced. (3) The grassland NPP showed spatial and temporal differences. The average NPP of grassland increased by 21.30, 16.47, and 36.17% during 1985–1995, 2005–2015, and 1985–2015, respectively. From 1995 to 2005, the average NPP decreased by 5.05 gC/m^2, which is equivalent to -3.61% of the average NPP in 1995. The total amount of grassland NPP in the study area was the greatest in 2015, showing increments of 36.37% and 16.61% compared with those in 1985 and 2005, respectively.

Keywords Grassland landscape index · Grassland productivity · Land-use/cover change · Net primary productivity (NPP)

5.1 Introduction

As one of the most widely distributed land-cover types on land, grassland accounts for approximately 20% of the global terrestrial surface area (Scurlock 2010). Its net primary productivity (NPP) accounts for 16% of the total global terrestrial ecosystem NPP (Conant et al. 2001; Zhou et al. 2010). China's grassland area is approximately 4 million km^2, accounting for 41.7% of China's total land area and 6–8% of the global grassland area (Ren et al. 2008). The grassland carbon reserves in China accounts for 9–16% (Ni 2002) of the global grassland carbon stocks. Grassland resources play an important role in China's ecological and environmental protection and socioeconomic development and are considered one of the important renewable natural resources (CCICCD 1996; Han et al. 2008; Nan 2005). Grassland provides

© Springer Nature Singapore Pte Ltd. 2020
W. Zhou et al., *Remote Sensing Monitoring and Evaluation of Degraded Grassland in China*, Springer Geography,
https://doi.org/10.1007/978-981-32-9382-3_5

vital materials, such as fodder, meat, milk, skin, wool, and other livestock products, for the economic development of animal husbandry (Kang et al. 2007). This land type also serves as significant habitat and evolution site for animals and plants in grassland ecosystems (Kang et al. 2007; Zhou et al. 2004). Sexual protection, wind and sand fixation, conservation of water resources, and carbon recycling are some of the important ecological functions of grassland, which is also the material carrier of China's multiethnic culture (Han et al. 2004).

China's grassland ecosystem and ecological environment have changed due to the intensification of global climate change with the main features of global warming, the adverse effects of extreme climate events, and the unreasonable and prolonged use of grassland resources by humans (Han et al. 2004). Desertification occurred, salinization accelerated, and the grassland ecological function decreased. Some studies have found that 90% of China's available natural grassland has experienced various degrees of degradation. Approximately 27.3% of China's natural grassland is affected by desertification, and about 400 million production and livelihood are affected by desertification. Furthermore, the economic loss caused by sandstorms can reach 54 billion yuan (Akiyama and Kawamura 2007).

Grassland degradation has a serious impact on ecological and environmental protection and socioeconomic development (Cao 2011; Liu et al. 2008). The degeneration of grassland in northern China and the frequent severe dust storms have received extensive attention from the international community. Since 1999, the Chinese government has implemented ecological restoration measures. This program has led to changes in land use and cover (LUC) and grassland coverage and ecological functions. In addition, climate change also affects the global ecological environment at an unprecedented rate. In some parts of eastern Asia, the frequency of extreme climatic events has increased by 5% (Qi et al. 2012a, b) during the past 30 years compared with that during 1961–2000. The temperature in China increased from 1950 to 2010 at the speed of 0.17 °C/10 year (Zhai et al. 2005), which is 0.13 °C/10 year higher than the rate of global temperature rise.

Affected by global climate change and human activities, China's grassland ecosystem has undergone significant changes. The main distribution areas of grasslands are found in the arid and semi-arid regions of northern China and the alpine grassland on the Tibetan Plateau. These locations are particularly sensitive to climate change and human disturbance. Therefore, studying the response of grassland dynamics to humans and climate change in these regions is important to understand global changes. The changes in grassland ecological environment have important theoretical significance. The change in landscape pattern and productivity in grassland coverage area is the most intuitive response of grassland ecosystem to climate change and land-use and cover change (LUCC). By using LUCC data, remote sensing, climate data, and nature, society, and economy data, this paper analyzed the changes in the grassland cover area, landscape index (LSI), and NPP of the study area from 1985 to 2015. The effect of climate and human activities on grassland ecosystem changes will provide a theoretical basis for the sustainable development of China's grassland ecosystem in the context of global change.

5.2 Methodology

5.2.1 Study Area

The study area in this paper is China's main pastoral area ($26°50'$–$53°23'$N, $73°40'$–$126°04'$E), which includes three provinces (Gansu Province, Shaanxi Province, and Qinghai Province) and four autonomous regions (Inner Mongolia Autonomous Region, Ningxia Hui Autonomous Region, Xinjiang Uygur Autonomous Region, Tibet Autonomous Region). The study area has a total area of 5.5 million km^2, and the climate is mainly arid and semi-arid. Accounting for 57.21% of the total land area of China, the grassland here has been used for a long period of time for grazing, but it has been destroyed in recent years, the dual effects of human and climate have resulted in the improvement of grassland destruction, and the implementation of projects such as returning grazing land to forests and grasslands has also been continuously reduced.

5.2.2 Land-Use Data

To analyze the spatial and temporal patterns of land-use changes across China, the Chinese Academy of Sciences built a data platform supported by the National Resources and Environment Database (NRED) in the late 1990s. Land-use datasets for 1985, 1995, 2005, and 2015 with a mapping scale of 1:100,000 were originally derived from Landsat images of corresponding years, and then a 1-km raster database was generated. According to the land-use classification system for the NRED dataset, the land use was categorized into six types: cropland, forest, grassland, waterbodies, built-up land, and bare land including desert. In this study, the land-use data for 1985, 1995, 2005, and 2015 with a 1-km resolution were downloaded from Data Sharing Infrastructure of Earth System Science (http://wdcrre.geodata.cn/, accessed 20 May 2017) (Liu et al. 2003).

5.2.3 Landscape Metrics

This article uses the Patch Analyst extension module in ArcGIS to calculate the landscape index of grassland in seven major pastoral areas in four periods. This module embeds some FRAGSTATS functions into ArcGIS software to facilitate the implementation of patches and types in ArcGIS and calculate the landscape indices of type and landscape scales (Yu et al. 2011). The definition and detailed description of various landscape indices are similar to those in FRAGSTATS (Wang et al. 2008). The following six indices were selected on the type scale: number of patches (NP), mean patch size (MPS), edge density (ED), area weighted mean shape index (AWMSI), area weighted mean fractal dimension (AWMFD), and interspersion juxtaposition index (IJI). The following two indices were selected on landscape scale: Shannon's diversity index (SHDI) and Shannon's evenness index (SHEI). The calculation process is implemented in ArcGIS 10.2.

5.2.4   CASA Model to Estimate NPP

Three models, namely parameter, statistical, and process models, are used to estimate terrestrial ecosystem NPP. Basing on the existing CASA model (Liu et al. 2003) and the land-use/land-cover classification standard of the Chinese Academy of Sciences, this paper examines the grassland NPP in the study area. This model mainly uses meteorological data and vegetation NDVI data as basic parameters. Effective radiation, temperature stress coefficient, water stress coefficient, and maximum light energy utilization efficiency of photosynthesis for vegetation were combined to account for vegetation NPP (Zhong et al. 2005; Han et al. 2004). The vegetation NPP is mainly determined by the photosynthetically active radiation (APAR) and light energy utilization (E) absorbed by the vegetation.

$$\text{NPP}(x, t) = \text{APAR}(x, t) \times E(x, t) \tag{1}$$

where NPP(x, t) represents the photosynthetically active radiation (PAR) absorbed by pixel x in month t, in units of MJ/m^2/month; and $E(x, t)$ represents the actual light energy utilization of pixel x in month t, in units: g C/MJ.

$$\text{APAR}(x, t) = \text{SOL}(x, t) \times \text{FPAR}(x, t) \times 0.5 \tag{2}$$

In the formula, the amount of total solar radiation absorbed by the pixel x in month t is expressed in units of MJ/m^2/month. The ratio of the absorption of the effective photosynthetically active radiation by the vegetation layer which the constant value of 0.5 indicates that the solar radiation available to the vegetation accounts for the total solar radiation (proportion of radiation).

Light energy utilization refers to the efficiency of the vegetation to convert the absorbed PAR into organic carbon. Under ideal conditions, vegetation has the maximum light energy utilization rate. However, in actual conditions, the true maximum light energy utilization rate (E) is also affected by temperature and precipitation. The specific formula is as follows.

$$E(x, t) = T_{E1}(x, t) \times T_{E2}(x, t) \times W_E(x, t) \times E_{\max}, \tag{3}$$

where $T_{E1}(x, t)$ refers to the stress effect of high temperature on light utilization (no unit), $T_{E2}(x, t)$ refers to the effect of low temperature on the utilization of light energy (no unit), $W_E(x, t)$ is the moisture coercion coefficient (no unit) that mainly indicates the degree of influence of the calculated NPP on water status, and E_{max} refers to the maximum light energy utilization under ideal conditions (unit: g C/MJ).

Temperature, moisture, and other stress factors are calculated according to previous methods. The maximum light energy utilization rate varies with vegetation types. This paper refers to the maximum light energy utilization rate of typical vegetation in China as simulated by Zhu (Zhu et al. 2006). Vegetation types include cropland, grassland, waterbodies, built-up land, and bare land, and the light energy utilization rate is 0.542 g C/MJ. The globally recognized value of 0.389 g C/MJ is selected as the maximum light energy utilization rate of the forest land.

5.2.5 Model Accuracy Verification

In this paper, 51 grassland biomass samples measured in the main pastoral areas of China in July and August of 2008 were used to translate the aboveground and underground biomass allocation ratios of Inner Mongolia grassland into grassland and underground vegetation according to Ma Wenhong et al. Productivity, which was evaluated from the NPP data measured in the grass, was used to verify the accuracy of the model. The following figure (Fig. 5.1) shows the correlation analysis between simulated and measured values of grassland NPP model in 2005, $R^2 = 0.7103$ ($P < 0.001$). The results show that the CASA model has high simulation accuracy and can be applied to the simulation of grassland NPP in the study area.

Fig. 5.1 Relationship between estimated net primary productivity (NPP) and observed NPP for grasslands in August of 2008

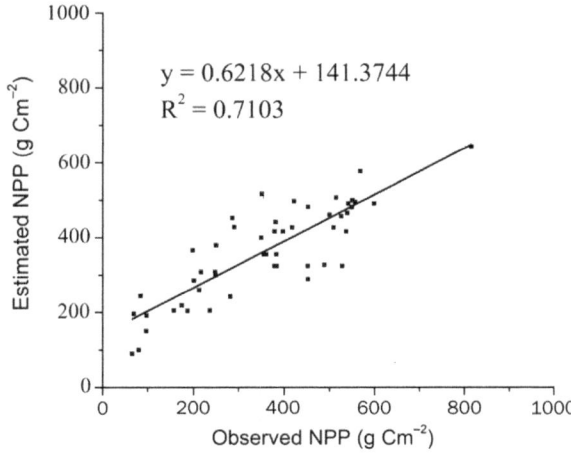

$$y = 0.6218x + 141.3744$$
$$R^2 = 0.7103$$

5.3 Implementations and Discussions

5.3.1 Analysis of Spatial and Temporal Dynamic Changes of Grassland

The land cover of the seven pastoral areas in 1985, 1995, 2005, and 2015 is shown in Fig. 5.2, and the dynamic changes in the grassland study area are shown in Tables 5.1 and 5.2. The sizes of the grassland study area in 1985, 1995, 2005, and 2015 were 248.34, 243.93, 245.80, and 2.4466 million km², respectively. The grassland areas in different pastoral areas changed differently. The grassland area in Tibet Autonomous Region was the largest in 1985 and accounted for 70.05% of the total land area in the region. The grassland in Xinjiang Uygur Autonomous Region accounted for 29.61% of the total land area in 1985 and 1995. The grassland areas in Inner Mongolia Autonomous Region, Qinghai, and Shaanxi were the largest in 1995 and accounted

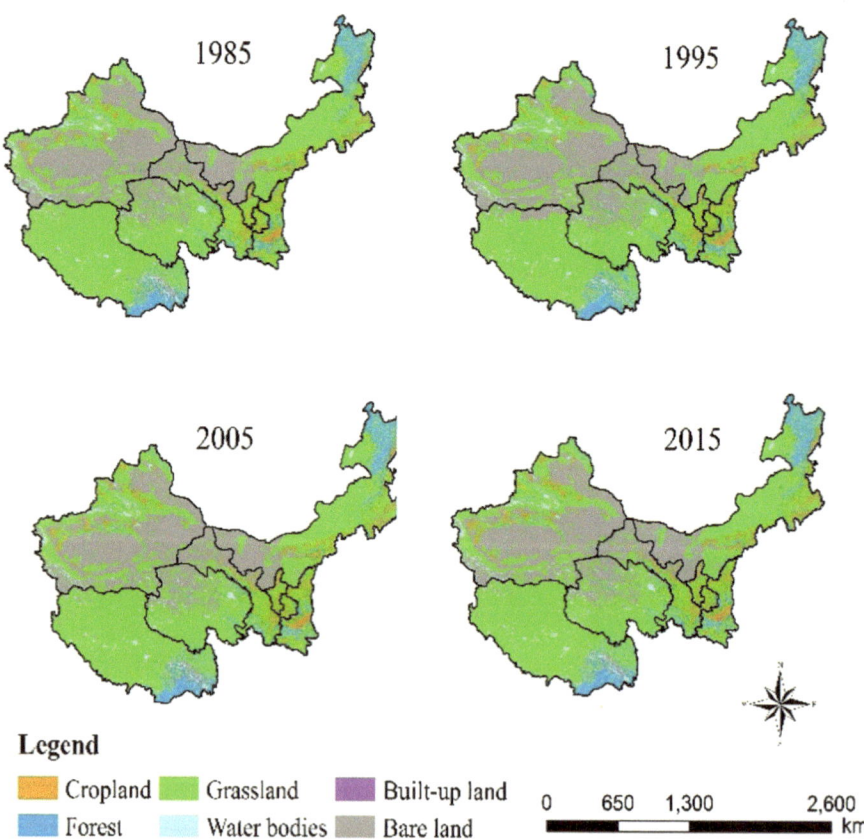

Fig. 5.2 Spatial distribution of grassland areas changes during 1985–2015

Table 5.1 Grassland annual mean NPP changes during 1985–2015/(g C/m^2)

Sections	Study area	Tibet	Inner Mongolia	Xinjiang	Qinghai	Gansu	Shanxi	Ningxia
1985	130.45	90.34	168.46	105.26	178.12	159.85	234.4	124.41
1995	164.35	119.56	199.26	137.29	193.56	217.71	281.0	143.53
2005	161.37	135.63	164.37	127.49	211.80	218.59	292.4	118.81
2015	184.53	145.89	200.32	136.73	238.30	276.22	368.3	164.24
Percentage change (%)								
1985–1995	25.99	32.34	18.28	30.43	8.67	36.19	19.89	15.37
1995–2005	−1.81	13.44	−17.51	−7.14	9.42	0.40	4.07	−17.22
2005–2015	14.35	7.57	21.88	7.25	12.51	26.36	25.95	38.24
1985–2015	41.46	61.49	18.91	29.90	33.79	72.80	57.15	32.01

Table 5.2 Grassland total NPP variations during 1985–2015/(Tg C)

Sections	Study area	Tibet	Inner Mongolia	Xinjiang	Qinghai	Gansu	Shanxi	Ningxia
1985	322.34	75.77	90.1	50.65	24.65	60.19	17.77	3.21
1995	398.95	90.64	113.34	66.06	73.34	30.01	22.05	3.47
2005	395.08	113.65	86.16	60.02	79.63	30.31	22.5	2.84
2015	452.16	122.4	105.2	64.36	89.56	38.36	28.34	3.95
Percentage change (%)								
1985–1995	23.77	19.63	25.8	30.44	197.54	−50.15	24.09	23.77
1995–2005	−0.97	25.4	−23.98	−9.14	8.57	1	2.04	−0.97
2005–2015	14.45	7.69	22.1	7.22	12.48	26.59	25.95	14.45
1985–2015	40.28	61.55	16.76	27.07	263.34	−36.26	59.48	40.28

for 50.03%, 52.96%, and 38.72% of the total land areas, respectively. The grassland area in Gansu reached the maximum in 2015 and accounted for 34.61% of the whole land area.

From 1985 to 1995, the grassland in the study area decreased by 18.92% (468,829 km^2). The reduction in grassland is mainly due to the mutual transformation of grassland to bare land or to forest, resulting in a decrease of 215,888 or 108,343 km^2, respectively. The conversion of bare land to grassland led to a net increase of 320,574 km^2 in grassland. In the spatial distribution, the area of grassland reduction is mainly distributed in Tibet, Gansu, and Ningxia. For these regions, the transfer of grassland to bare land is the main reason for the decrease in grassland area. The net conversion of grassland to bare land in the three provinces reached to 56,161, 14,211, and 1705 km^2, respectively. By contrast, the increase in grassland area was distributed in the other three provinces, of which the highest was in Qinghai grassland area at 90,794 km^2. In these regions, 66,377 km^2 of grassland was transformed from bare land and 14,562 km^2 was from forests. The grassland area in Inner Mongolia increased by 72,846 km^2, of which the net area transformed from bare land to grasslands reached to 23,607 km^2. In Xinjiang, the transfer was only 13 km^2 from 1985 to 1995.

From 1995 to 2005, the grassland in the study area increased by 18,694 km^2, equivalent to 0.77% in 1995. The increase of grassland is mainly from the bare land to grassland, covering an 33,8709 km^2, while the reduction of grassland is mainly converted to bare land and forest 291,758 and 1,133,956 km^2. In the spatial distribution, newly added grassland is concentrated in Tibet and Gansu, 10.63% and 0.70% of the grassland area in 1995. According to the statistical analysis, there were 80,586 km^2 bare lands transferred to grassland in Tibet, while the number converted to bare land was 105,028 and finally the net increase of grassland area was 83,862 km^2. There were 23,307 km^2 grassland transferred from the cropland in Gansu Province. It is widely distributed in the other five provinces, among which, the conversion of grassland and bare land, farmland, and forest in Inner Mongolia leads to the largest reduction of grassland, which is 65,570, 26,379, and 25,350 km^2, respectively. There was 50,834 km^2 (in Xinjiang) and 76,285 km^2 (in Qinghai) grassland caused by the conversion to the bare land. The conversion of Ningxia grassland to cropland; and bare land led to a net loss of 5977 and 1570 km^2 respectively, but at the same time, 5265 km^2 of cropland was transferred to grassland. Meanwhile, the net reduction of grassland to forest and bare land in Shaanxi was 1016 and 648 km^2.

From 2005 to 2015, the grassland area in the study area decreased by 11,414 km^2, equivalent to 0.46% of the grassland in 2005. The decrease of grassland is mainly the mutual transfer of grassland and cropland, with a total reduction of 9150 km^2. The increase of grassland was mainly from cropland and bare land, resulting in 5845 km^2. For pastoral areas, the decrease of grassland area concentrated in Xinjiang, Shaanxi, and Ningxia, and the reduction rates were 17.37%, 1.797%, and 1.92%, respectively. The conversion of grassland to bare land in Xinjiang and Tibet was 153 km^2 and 264 km^2, respectively, while the net increase of 245 and 221 km^2 in Gansu and Ningxia Hui Autonomous Region was from bare land. Since the conversion from grassland to cropland and bare land happened, more pastoral grassland has reduced, and it was 821 km^2 in Inner Mongolia, 209 km^2 in Ningxia, 115 km^2 in Shaanxi Province, and 265 km^2 to grassland at the same time (Figs. 5.3, 5.4 and 5.5).

5.3.2 Changes in Pattern of Grassland Landscapes

The SHDI and SHEI in the whole study area increased slowly, whereas the landscape heterogeneity increased. Among the seven pastoral areas, the SHDI and SHEI in Xinjiang, Gansu, Shaanxi, and Ningxia were decreased. However, the SHDI in Qinghai Province increased from 2005 to 2015, whereas the SHEI was relatively flat. This finding showed that the landscape pattern in Qinghai Province was complex, and the spatial distribution of land-cover/land-use types was uneven. The SHDI and

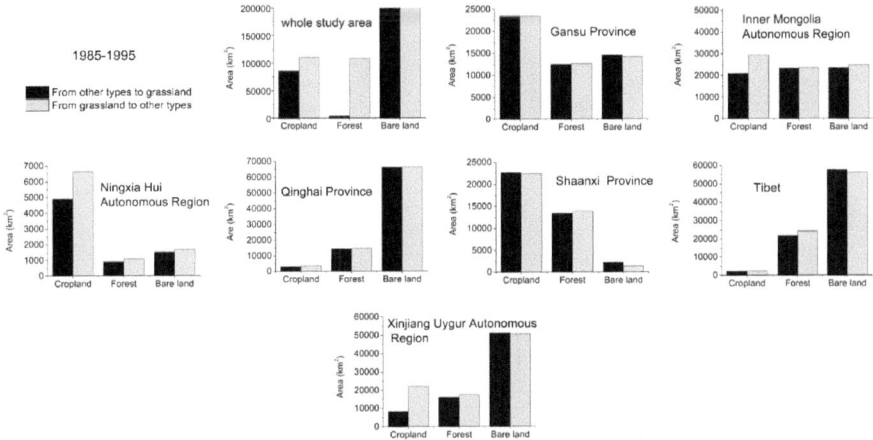

Fig. 5.3 Conversion between grassland and other land-cover types during 1985–1995

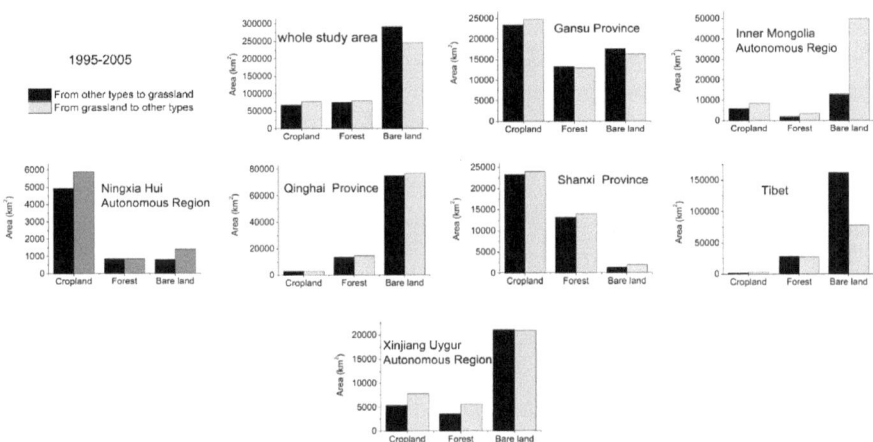

Fig. 5.4 Conversion between grassland and other land-cover types during 1995–2005

SHEI in Tibet show an increasing trend from 1985 to 1995 but a decreasing trend from 1995 to 2015. This result showed that the landscape pattern in Tibet was simple, and the space of each land-cover type was evenly distributed.

In the study area, the number of patches was the smallest in 1985 and 1995, whereas the mean patch size value was the largest. The change indicated that the degree of fragmentation and spatial heterogeneity of grassland was low in 1995. The NP and MPS variations in different provinces were inconsistent. In 2015, the NP in Inner Mongolia and Gansu was the lowest, whereas the MPS was the highest, indicating that the pasture fragmentation and spatial heterogeneity of the two pastoral areas were the lowest in 1995. Inner Mongolia, Gansu, Qinghai, and Ningxia reached

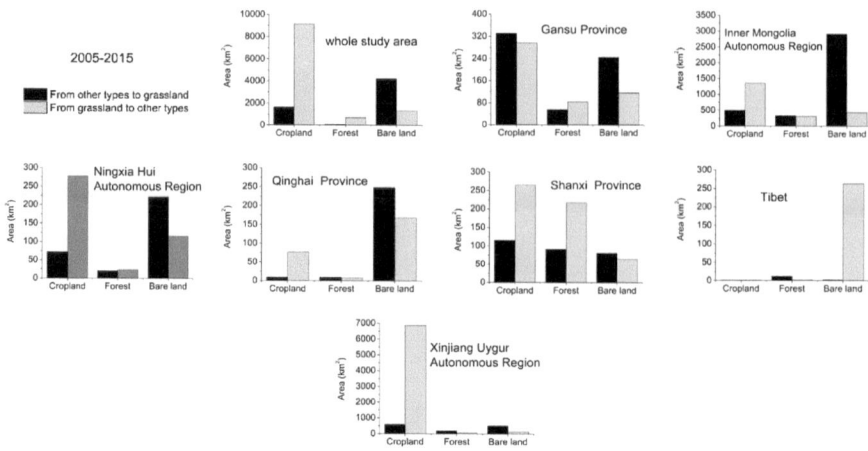

Fig. 5.5 Conversion between grassland and other land-cover types during 2005–2015

the minimum NP and maximum MPS in 2015. However, Xinjiang had the smallest NP and highest MPS in 1985 and 1995, revealing that grass fragmentation and spatial heterogeneity were the lowest during these years.

Except for Ningxia, no significant difference in the ED of grassland vegetation was found between 2005 and 2015, indicating no significant change in grass fragmentation and marginal effect. The AWMSI at the regional scale was high in 1985 and 1995, but low in 2005 and 2015. Hence, the grass patch shape in the whole area was simplified and regularized from 1995 to 2005. The AWMSI in Qinghai, Inner Mongolia, and Gansu showed basically the same rules of variation with the whole study area. However, the AWMSI increased sharply in Ningxia in 2005–2015, which indicates that the patch shape of grassland in the province was complicated, and the edge effect increased (Fig. 5.6).

The AWMFD in the entire study area did not change in many years. The grassland landscape pattern was not affected by human activities. The trends of AWMFD in Inner Mongolia, Tibet, and Qinghai are similar to those in the entire study area. However, Tibet has the largest AWMFD, which indirectly reflects that the grassland landscape pattern in this province is less affected by human activities. This phenomenon is also caused by the unique geographical location of Tibet. The AWMFD in Gansu and Shaanxi was the highest in 1995 and 2005 but was low in 1985 and 2015, indicating that the effect of human interference on grassland landscape pattern was less in 1985–1995 but increased in 2005–2015 and 1995. In 2005, less interference was observed from humans. In Xinjiang, AWMFD remained unchanged from 1995 to 2005 but increased sharply from 2005 to 2015, indicating that grassland in Xinjiang was reduced by human interference. In addition, IJI increased throughout the study area and in the seven provinces and regions between 2005 and 2015. The connectivity and cohesion of grass patches decreased, but IJI declined significantly

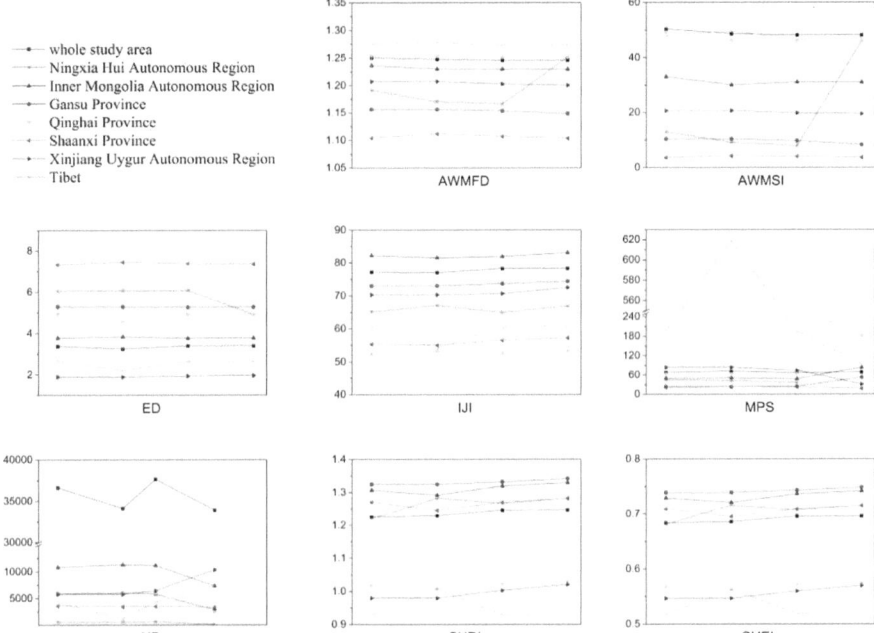

Fig. 5.6 Changes in landscape index for the whole study area and the seven provinces during 1985–2015. *Note* AWMFD: area weighted mean fractal dimension, AWMSI: area weighted mean shape index, ED: edge density, IJI: interspersion juxtaposition index, MPS: mean patch size, NP: number of patches, SHDI: Shannon's diversity index, SHEI: Shannon's evenness index

in Tibet during 1985–1995 and in Ningxia during 1995–2005. This finding shows that the patch connectivity and cohesion of grassland are greatly improved.

5.3.3 Changes in NPP of Grassland

The spatial distribution of average grassland NPP shows evident heterogeneity. The areas with high NPP values are distributed in the northeastern Inner Mongolia, Xinjiang Tianshan, and the Al Taishan region southern Shaanxi, southern Gansu, the source of three rivers, Qilian Mountains, the southeast of Tibet, northwest of Tibet, and the southern of Xinjiang the grassland NPP values are the lowest. The average NPP of grassland in the entire study area increased by 25.99%, 14.35%, and 41.46% from 1985 to 1995, 2005 to 2015, and 1985 to 2015, respectively. The average NPP in 1995–2005 was reduced by 2.98 gC/m^2, which is equivalent to 1.81% of the average NPP in 1995. The changes in various provinces are as follows. The average NPP of grassland in Xinjiang slightly changed from 1995 to 2005 and then increased by 32.03 and 9.24 gC/m^2 in 1985–1995 and 2005–2015, respectively. The average NPP in Tibet was obtained in 1995–2005. The value was reduced by 34.89 gC/m^2 but increased by 30.80 and 35.96 gC/m^2 in 1985–1995 and 2005–2015, respectively. The average NPP change of grassland in Qinghai province increased by 15.44 and 18.24 gC/m^2 frame 1985 to 1995 and 1995 to 2005. The increase from 2005 to 2015 was 26.50 gC/m^2. The average NPP in Inner Mongolia grassland decreased by 34.89 gC/m^2 from 1995 to 2005, which is equivalent to the value in 1995 that is 17.51% of grassland NPP. The grassland NPP in Gansu Province changed slightly from 1985 to 2005 but increased by 57.63gC/m^2 from 2005 to 2015. Shaanxi Province showed an increasing trend of NPP from 1985 to 2015 with a total increase in 30 years. At 133.95 gC/m^2, NPP in grassland decreased from 1995 to 2005, with a decrease of −24.72 gC/m^2, and an increase of 45.43 gC/m2 from 2005 to 2015, which was equivalent to 38.24% of NPP in 2005 (Fig. 5.7).

The total NPP in grassland in the study area was the highest in 2015, which was 452.16 Tg C. This value was 40.28%, 13.24%, and 14.45% higher than that in 1985, 1995, and 2005, respectively. The changes in different provinces and regions are quite different. The total grassland NPP in Xinjiang increased by 30.44% from 1985 to 1995 but decreased by 9.14% in 2005. The total grassland NPP in Tibet showed an increasing trend at each stage from 1985 to 2015. When the average NPP value

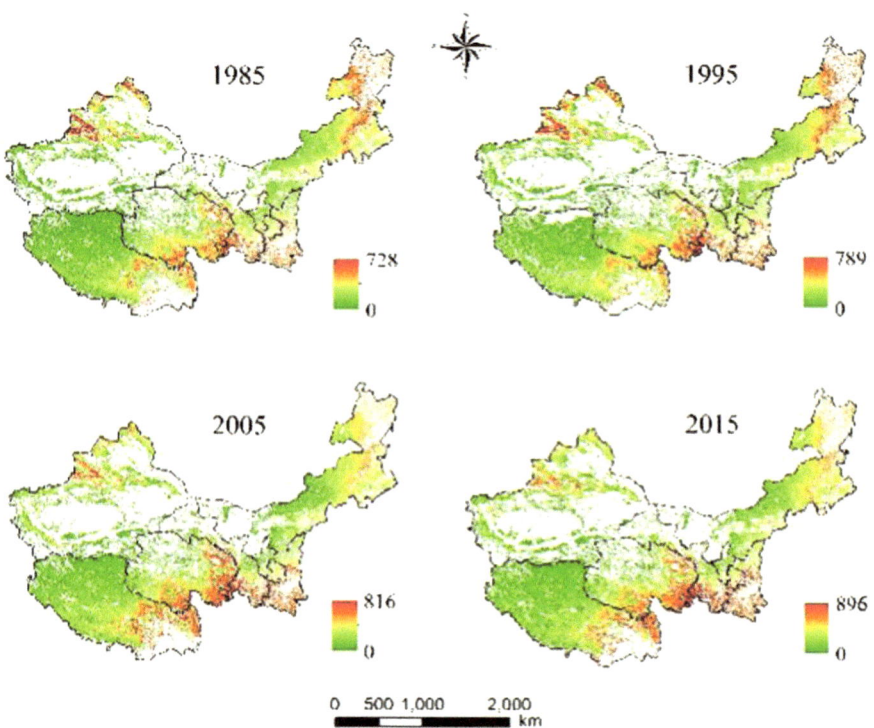

Fig. 5.7   Spatial distribution of grassland mean NPP during 1985–2015

increased by 61.49% in 2015, the total NPP value increased by 61.55%, though the grassland area in Tibet decreased by 0.17% in 1985–2015. The total NPP in grassland in Qinghai Province has increased from 1985 to 2015, especially in 1985–1995 with an increase of 197.54%. When the average NPP of grassland increased from 2005 to 2015, the total NPP also increased. The total amount of NPP in Inner Mongolia in 2005 was −23.98% lower than that in 1995, because the grassland area and average NPP were reduced by −8.05% and 17.51%, respectively, whereas the total amount of NPP increased by 22.10% from 2005 to 2015. The total amount of grassland NPP in Gansu Province decreased by −50.15% between 1985 and 1995 but increased by 26.59% between 2005 and 2015. The change rule of total NPP in Shaanxi grassland is consistent with the change in average NPP. The value showed an increasing trend from 1985–2015 with a total increase of 10.57 Tg C. The total NPP in Ningxia grassland was the highest in 2015 with an increase of 38.73% compared with that in 2005.

5.3.4 Comparisons

(1) **Influence of climate change on landscape pattern and grassland productivity**

Climate change is the main factor for the inter-annual change of terrestrial vegetation activity (Keeling et al. 1996; Weltzin et al. 2003). It mainly affects the growth of grassland vegetation by the changes in temperature and precipitation. The increasing precipitation in arid regions is conducive to the growth of forage grass (Yang et al. 2008). In contrast, the increase in temperature also increases vegetation and soil evapotranspiration but reduces available soil moisture. This phenomenon makes the environment arid and not conducive to forage growth (Shen et al. 2012). In the context of global climate change, China's climate has also undergone corresponding changes. To analyze the climate change patterns in the study area, this paper collected temperature and precipitation data from 1961 to 2010. The trend of changes and the regularity of anomalies were also evaluated.

The annual temperature in the study area has gradually increased from 1961 to 2010, with an increase rate of 0.39 °C/10 year, which is higher than the rate of annual temperature increment of 0.17 °C/10 year from 1951 to 2000 in China. The inter-annual rate in annual precipitation is 5.2 mm/10 year. The annual fluctuation of precipitation is relatively large, especially in 1998, due to the increment in floodwater (Zhu et al. 2006). The precipitation is abnormally high and was affected by strong El Niño in 1982. With the drought in north and south of China, the precipitation is lower than the annual average. Some studies reported that China's precipitation did not change significantly during 1951–2000. However, the precipitation in different regions was different (Shi et al. 2007). The regions where precipitation increased significantly were located in northern China, the northwestern arid regions, and the Tibetan Plateau. A previous report is basically consistent with the conclusions in the current paper that the temperature in the study area has continued to increase during the past 50 years, and the precipitation has not changed significantly. However, since 1980s, the climate in the arid northwestern China has transformed from warm to warm and humid. Significant differences, such as the trend of warming and drying in Inner Mongolia from 1951 to 2010, are observed in the precipitation changes in different regions (Liu et al. 2012; Lu et al. 2009).

During 1981–2010, precipitation increased, and in 20 years, the 20-year precipitation anomaly was positive. The precipitation increased from 318 mm in 1980s to 324 mm in 2000–2010, an increase of 2%. However, the temperature continued to increase from 5.5 °C in 1980s to 6.75 °C in 2000–2010, an increase of 1.25 °C. Some studies have suggested that with a temperature increase of 1 °C, the annual potential evapotranspiration (PET) increases by 5.25% +1.55% (Le Houérou 1996). Therefore, the increase in precipitation is not sufficient to offset the increase in temperature, resulting in increased drought, vegetation, and evapotranspiration. The reduction on soil moisture and such climatic conditions are the basis for desertification. Coupled with the interference of human activities, such as overgrazing and indiscriminate digging, these phenomena lead to grassland desertification and ultimately to the original vulnerability of grassland to desert transformation. During 1985–1995 and 2005–2015, large areas of grassland were migrated to the desert. A small portion of deserts shifted to grassland during 1995–2005. This migration was mainly due to the 10.75% increase in grassland area in Tibet during the period and was the reason for the increase in grassland area in the district (Akiyama and Kawamura 2007; Teague and Dowhower 2003; Zha 2004). Although human activities have always been considered as the dominant driving factor in grassland degradation, climate aridity accelerates the development of grassland degradation to a certain extent, especially in the arid–semi-arid regions of northwestern China. The continuous increase in temperature from 1961 to 2010 promotes the conversion of grassland to desert at some extent (Han et al. 2008). In addition, studies in the Hunshandake Sandy Land have also confirmed that climate change plays an important role in desertification (Yang 2010) (Fig. 5.8).

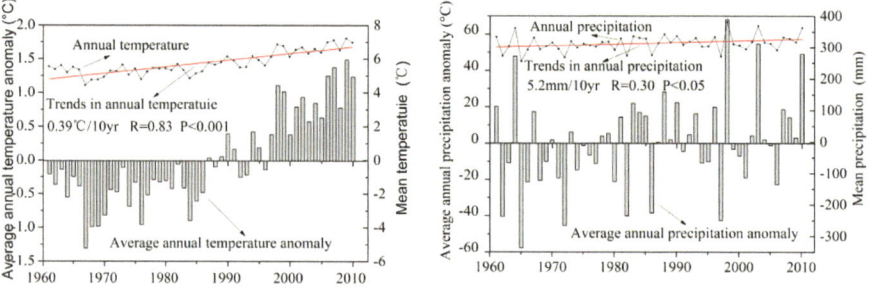

Fig. 5.8 Inter-annual changes of the mean temperature, precipitation, and anomaly during 1961–2010

(2) **Effects of human factors on grassland coverage area and productivity**

① **Effect of overgrazing**

Overloaded grazing is the main driving force for grassland degradation in China and is particularly prominent in grassland degradation in northern China (Akiyama and Kawamura 2007). The degradation of grassland in China began in the 1960s. Since then, the degraded area of grassland in China has increased at a rate of 15%/10 year (Wang et al. 2005). In the past 10 years, the degradation rate of grassland in China has increased from 55 to 90% (Du 2006). The rapid increase in population and economic development has led to a sharp increase in the demand for animal husbandry products and eventually to a rapid increase in the number of grazing animals. The number of grazing animals in China increased from 11.9 million in 1950s to 61.3 million in 2001 based on the statistics found in China. The total number of livestock showed a rapid increase from 1978 to 2011 (Fig. 5.10), though the number of livestock has decreased since 2003. Overgrazing occurs on a large area of grassland, and the grassland can no longer recuperated under the constant stamping and feeding disturbances. In some arid areas, grassland overloading rate has reached 50–120% or even 300%. As shown in Fig. 5.9, overloaded grazing occurred in all six provinces in northwest China, except for Shaanxi Province. The number of livestock in Xilin Gol League in Inner Mongolia increased from 2 million in 1977 to 18 million in 2000, causing about one-third of grassland degradation. As of 1995, China's degraded grassland area accounted for 50.2% of China's available grassland area. As of 1999, 32% of grassland in Inner Mongolia was overloaded with grazing, whereas 60% had varying degrees of degradation (Du 2006; Zhu 1997).

Under continuous overgrazing, animal feeding and stamping will reduce grassland vegetation, coverage, and productivity. At the same time, animal stamping will change the soil structure, increase soil hardness, reduce the effective soil moisture, and expose patches on the surface. This phenomenon provides conditions for wind erosion desertification (Zhu 1997). The results indicate that, except for the increase in grassland area between 1995 and 2005, the grassland area showed a decreasing trend

Fig. 5.9 Annual changing trend of livestock numbers in China during 1978–2011

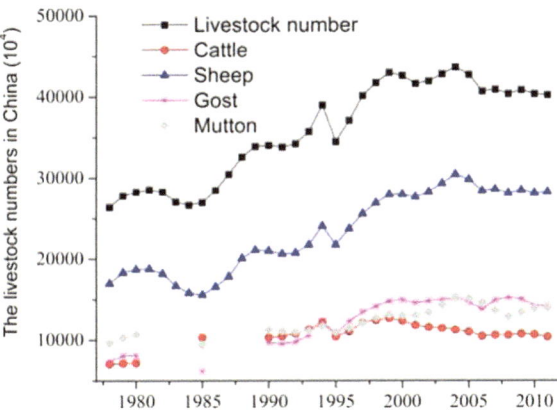

Fig. 5.10 Grassland
overgrazing situation in main
animal husbandry provinces

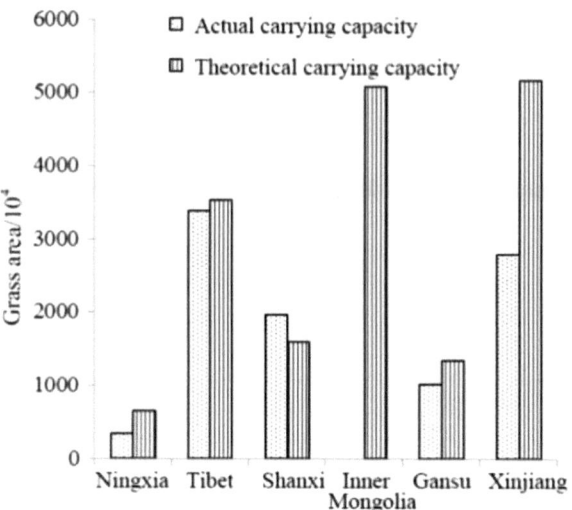

that is mainly due to the shift of grassland to desert and farmland. Approximately 468,800 km² of grassland was converted to other lands during 1985–2015. For the utilization types, 46.05% were converted to desert, and 23.42% were converted to farmland, which ultimately led to a net decrease of 1.48% in grassland area.

② **Influence of grassland reclamation**

Rapid population growth has rapidly increased food demand, and grassland has been reclaimed as farmland on a large scale (Jiang et al. 2006). Under the guidance of the "grain as the key" policy, a large area of grassland has rapidly undergone sandy desertification. From 1949 to 1999, 193,000 km² of grassland was reclaimed as farmland, and 18.2% of China's increased farmland area originated from grassland reclamation. Between 1958 and 1976, 20,000 km² of grassland was reclaimed into farmland in Inner Mongolia (Chuluun and Ojima 2002). This shift took place mainly in the agro-pastoral ecotone. During urbanization, grasslands around the city were reclaimed to meet the food demand. This study finds that from 1985 to 2015, 109,800 km² of grassland was reclaimed as farmland, whereas 85,700 km² of farmland was converted into grassland. This mutual conversion between farmland and grassland did not significantly reduce the total grass area. However, the overall quality of the grassland was reduced during conversion. Considering that most of the grasslands that were reclaimed as farmland belong to high-quality pasture grasslands, the grasslands transformed from farmland are mostly abandoned lands or land with low productivity. This kind of grassland reclamation often ends in failure, especially in the dry grassland area. Approximately 30–80% of the reclamation farmland gradually degenerates and is finally abandoned. After the grassland has been reclaimed, especially in a continuous arid environment, it will be degraded into desertified land with lost productivity and destroyed land resources after 30 years. Xinjiang had a net reduction of 97,428 km² in grassland from 1985 to 2015, which was equivalent to 20.20% in 1985. This phenomenon was mainly due to the continuous increase in temperature and glacial meltwater, which led to a drastic increase in grassland land reclamation and a large area of grassland degradation to desertification. The joint action of the people led to the reduction of grassland area.

The direct consequence of grassland degradation is the decline in grassland productivity. According to statistics, the current grass yield per unit area decreased by 30 to 50% compared with that in 1960s. Inner Mongolia grassland surveys showed that the aboveground NPP in Inner Mongolia grassland decreased by 53% (Qi et al. 2012a, b) from 1961 to 2010. This decrease is mainly due to overgrazing and climate change. The changes in temperature and precipitation over the past 30 years are shown in Fig. 5.11. The trend of aridity development in central and eastern Inner Mongolia and Shaanxi Province is clear. Therefore, the combination of overloaded grazing and climate drought reduced NPP in Inner Mongolia and Shaanxi Province during 1995–2005. The average NPP of grassland in the year was significantly low with a decrease of 20.04 and 2.03%. However, for the entire study area, the grassland NPP showed an increasing trend due to the significant increase in NPP in the grasslands of Xinjiang, Tibet, Qinghai, Gansu, and Ningxia. This increase may be due to increased precipitation in most parts of Xinjiang. In particular, the precipitation in Tianshan, Altai Mountains, southern Xinjiang oasis, southwestern Tibet, most parts of Qinghai, and northwestern Gansu generally increased. This phenomenon promoted the growth of pasture, improved grassland quality, and consequently increased NPP.

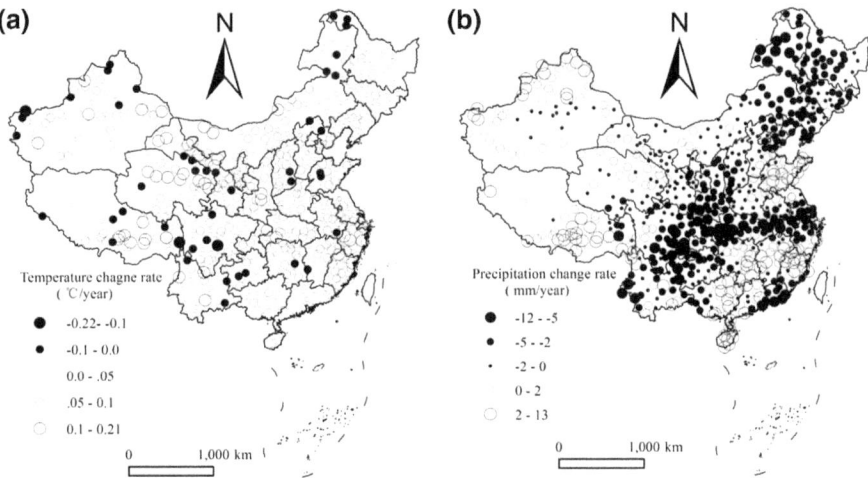

Fig. 5.11 Inter-annual changing trend of temperature and precipitation in China during 1982–2010

5.3.5 Effects of Ecological Restoration Measures on Grassland Productivity

The Chinese government's protection of grassland ecosystems began with the enactment of the Grassland Law in 1985. Later, the government established a series of laws and regulations concerning the protection of grassland resources and desertification. However, the deterioration of grassland in the north and the frequent occurrence of dust storms have recently pushed the Chinese government to put importance to the protection of grassland ecosystems and the restoration of degraded ecosystems (Zhang et al. 2012). Since 1999, the Chinese government has successively issued the project of returning farmland to forests and returning grazing land to grassland. The implementation of ecological restoration projects has played a positive role in the restoration of degraded grassland and the protection of grassland resources (Liu et al. 2008; Tong et al. 2004).

The project of returning farmland to forests and grass that began in 1999 was piloted in Gansu, Shaanxi, and Sichuan. By 2002, the project was expanded to 25 provinces and autonomous regions. This project mainly included the following restoration measures: conversion of degraded sloping farmlands into forests or grasslands, and closure of mountains and forests, and planting of artificial grass. This study found that the mutual transformation between farmland and grassland from 1985 to 2015 was conducive to an increase in the area of grassland, resulting in a net increase of 87,600 km^2 of grassland. The newly added grassland is distributed around the original grassland, which is beneficial to the improvement of the structure and function of the ecosystem. This procedure is also conducive to the conservation of biodiversity and the improvement of the stability of the ecosystem. Therefore, the conversion of farmland to grassland or forest is beneficial to ecosystem restoration.

The project of returning farmland to forests and grasslands may positively affect the reversal of desert to grassland in the study area from 2000 to 2010. As shown in Fig. 5.12a, the data show that the artificial grass planting in the study area increased rapidly from 2001 to 2003, reaching 1.72 million hectares in 2003 and accounting for 80% of the total area of artificial grass planted in China. Aircraft sowing and grass seed also showed improvement. The large increase in acreage and the combined effect of the three measures have increased the annual fluctuations in the area of newly added grass (Fig. 5.12b), reaching a maximum of 4.81 million hectares in 2006 and accounting for 60% of the national total; the former and newly added grass areas accounted for the largest proportion (50 and 58%) in the study area and dictated the trend of the two indicators. The artificial planting of grass and aircraft sowing will gradually expand the grassland area into the desert to achieve the purpose of "people entering the sand and receding." On the one hand, improved planting of grass species will make planting pasture suitable for the arid ecological environment. On the other hand, the excellent pasture can improve the pasture production. With the increase in grassland coverage, scattered grassland grows in patches, the degree of fragmentation decreases, and the dominance increases. Ultimately, the stability of grassland ecosystems increases, and productivity increases.

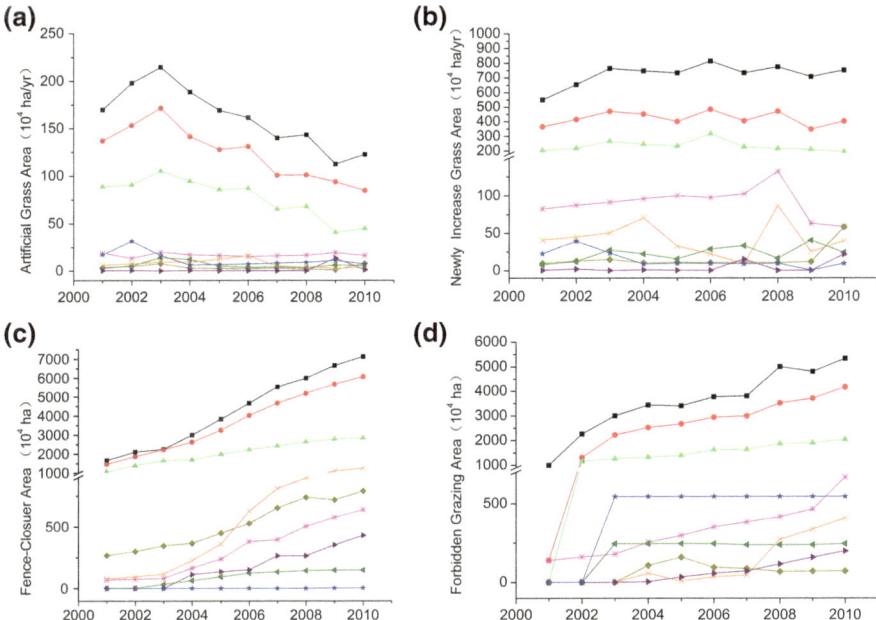

Fig. 5.12 Artificial grass area in China (**a**), the newly increase grass areas each year (**b**), the fence-closure area (**c**), and forbidden grazing area (**d**) during 2001–2010

To further promote the restoration of degraded grassland and the protection of grassland resources and improve the frequency of dust storms, the Chinese government began implementing the project of returning grazing to grassland in 2003. The grazing pressure was attenuated through measures, such as grazing bans, grazing and rotational grazing, and degrading. The grass is then restored. From 2001 to 2010, the grazing prohibition area increased yearly in all provinces and regions in the northwestern province. In 2010, the grazing prohibition area in the seven provinces and regions accounted for 78% of the country's area. The prohibition of grazing area in Inner Mongolia accounted for 38% of the grazing prohibited area in the country. A relatively high ban was placed on grazing in Shaanxi and Gansu Province (12.5, 10%). The area of fenced enclosures has increased significantly (Fig. 5.12d). The area enclosed by fences in the seven provinces and autonomous regions accounts for 85% of the total fenced area, and the fenced area in Inner Mongolia accounts for 40% of the country's total.

Grazing and fencing greatly reduce the pressure on grassland grazing. Without external interference, the biodiversity, stability, and productivity of grassland ecosystems can be recovered quickly (Bai et al. 2004), and long-term enclosed fences with large areas will increase the coverage of grassland. For a piece of degraded desert grassland in Alxa, the vegetation coverage has increased by 50%, and biomass has increased by 56% (Pei et al. 2008) after 6 years of prohibiting grazing. The number of years of closure increased, and grassland biomass and soil carbon density continued to increase (Jia et al. 2009). After the implementation of ecological restoration measures in Maqu County, Gansu Province, the grassland was significantly improved (Wang et al. 2009).

This study also found that the project of returning farmland to forests and grassland has the highest intensity in Inner Mongolia with evident ecological effects. This finding can be verified from the above-mentioned changes in artificial grass, fence enclosure and grazing prohibition areas, and grassland landscape index. From 1995 to 2005, SHDI and SHEI decreased, whereas ED, IJI, and AWMSI decreased. The change of landscape index showed that the dominance of Inner Mongolia grassland increased, the degree of fragmentation decreased, the cohesion of spatial distribution increased, and the shape of plaque became regular. The reason was as follows: The arid climate in 1995–2010 and the opening and overgrazing of grassland were still serious (Wang et al. 2004). After the implementation of the national ecological restoration measures, some desertification areas, such as Horqin Sandy Land and Mao Usu Sandy Land, have undergone a significant reversal of desertification. However, overall desertification is continuously intensifying. From 1986 to 2000, human activities played a decisive role in the development and reversal of desertification in Yulin, Shaanxi. In Xinjiang, Tibet and Qinghai, the relative proportion of grazing bans and fenced areas is relatively low (Sun et al. 2017). Therefore, the intensity of returning livestock to grassland, not just typical and severely degraded areas, must be increased.

5.4   Conclusions

(1)   The sizes of the grassland study area in 1985, 1995, 2005, and 2015 were 248.34, 243.93, 245.80, and 244.66 million km^2, respectively. From 1985 to 2015, the grassland was reduced by 36,800 km^2.

(2)   The heterogeneity of the landscape pattern increased during 1985–2015. The degrees of fragmentation and spatial heterogeneity of grassland were low in both 1985 and 1995. The shape of grass patches was simplistic and regular in 2005–2015.

(3)   The average NPP of grassland has significant regional differences in spatial distribution. The total size of grassland NPP in the study area was the lowest in 2005 but reached the greatest in 2015.

References

Akiyama T, Kawamura K (2007) Grassland degradation in China: methods of monitoring, management and restoration. Grassland Sci 53(1):1–17

Bai YF, Han XG, Wu JG, Chen ZZ, Li LH (2004) Ecosystem stability and compensatory effects in the Inner Mongolia grassland. Nature 431(7005):181–184

Cao SX (2011) Impact of China's large-scale ecological restoration program on the environment and society in arid and semiarid areas of China: achievements, problems, synthesis, and applications. Crit Rev Environ Sci Technol 41(4):317–335

CCICCD (1996) China National Committee for the Implementation of the UN Convention to Combat Desertification. China country paper to combat desertification. China Forestry Publishing House, Beijing

Chuluun T, Ojima D (2002) Land use change and carbon cycle in arid and semi-arid lands of East and Central Asia. Sci China Ser C Life Sci (English Edition) 45:48–54

Conant RT, Paustian K, Elliott ET (2001) Grassland management and conversion into grassland: effects on soil carbon. Ecol Appl 11(2):343–355

Du QL (2006) Sustainable development strategies of China prataculture. China Agric Press, Beijing, pp 58–59

Han Y, Han J, Zhang Y, Wang K (2004) Effects of transforming cropland into grassland on soil phosphorus and potassium in agro-pastoral transitional zone. J Soil Water Conserv 18(4):24–28

Han JG, Zhang YJ, Wang CJ, Bai WM, Wang YR, Han GD (2008) Rangeland degradation and restoration management in China. Rangeland J 30(2):233–239

Jia HT, Jiang PA, Zhao CY, Hu YK, Li Y (2009) Influence of fencing time on carbon distribution in grassland ecosystem. Agric Res Arid Areas 27(1):33–36

Jiang GM, Han XG, Wu JG (2006) Restoration and management of the Inner Mongolia grassland require a sustainable strategy. AMBIO J Hum Environ 35(5):269–270

Kang L, Han X, Zhang Z, Sun OJ, Kang L, Han X (2007) Grassland ecosystems in China: review of current knowledge and research advancement. Philos Trans R Soc B Biol Sci 362(1482):997–1008

Keeling CD, Chin J, Whorf TP (1996) Increased activity of northern vegetation inferred from atmospheric CO_2 measurements. Nature 382(6587):146–149

Le Houérou HN (1996) Climate change, drought and desertification. J Arid Environ 34(2):133–186

Liu J, Liu M, Zhuang D, Zhang Z, Deng X (2003) Study on spatial pattern of land-use change in China during 1995–2000. Sci China 46(4):373–384

Liu JG, Li SX, Ouyang ZY, Tam C, Chen XD (2008) Ecological and socioeconomic effects of China's policies for ecosystem services. Proc Nat Acad 105(28):9477–9482

Liu Y, Liu R, Chen JM (2012) Retrospective retrieval of long-term consistent global leaf area index (1981–2011) from combined AVHRR and MODIS data. J Geophys Res Biogeosci 117(G4):4003

Lu N, Wilske B, Ni J, John R, Chen J (2009) Climate change in Inner Mongolia from 1955 to 2005—trends at regional, biome and local scales. Environ Res Lett 4(4):45006

Nan ZB (2005) The grassland farming system and sustainable agricultural development in China. Grassland Sci 51(1):15–19

Ni J (2002) Carbon storage in grasslands of China. J Arid Environ 50(2):205–218

Pei SF, Fu H, Wan CG (2008) Changes in soil properties and vegetation following exclosure and grazing in degraded Alxa desert steppe of Inner Mongolia, China. Agr Ecosyst Environ 124(1–2):33–39

Qi JG, Chen JQ, Wan SQ, Ai LK (2012a) Understanding the coupled natural and human systems in dryland East Asia. Environ Res Lett 7(1):1–7

Qi JG, Chen JQ, Wan SQ, Ai LK (2012b) Understanding the coupled natural and human systems in dryland East Asia. Environ Res Lett 7(1):15202

Ren JZ, Hu ZZ, Zhao J, Zhang DG, Hou FJ, Lin HL (2008) A grassland classification system and its application in China. Rangeland J 30(2):199–209

Scurlock JMO (2010) The global carbon sink: a grassland perspective. Glob Change Biol 4(2):229–233

Shen WS, Li HD, Sun M, Jiang J (2012) Dynamics of aeolian sandy land in the Yarlung Zangbo River basin of Tibet, China from 1975 to 2008. Global Planet Change 86:37–44

Shi YF, Shen YP, Kang E, Li DL, Ding YJ, Zhang G (2007) Recent and future climate change in northwest China. Clim Change 80(3):379–393

Sun Q, Li B, Zhang T, Yuan Y, Gao X, Ge J (2017) An improved Biome-BGC model for estimating net primary productivity of alpine meadow on the Qinghai-Tibet Plateau. Ecol Model 350:55–68

Teague WR, Dowhower SL (2003) Patch dynamics under rotational and continuous grazing management in large, heterogeneous paddocks. J Arid Environ 53(2):211–229

Tong C, Wu J, Yong S, Yang J, Yong W (2004) A landscape-scale assessment of steppe degradation in the Xilin River Basin, Inner Mongolia, China. J Arid Environ 59(1):133–149

Wang ZY, Ding YH, He JH, Yu J (2004) An updating analysis of the climate change in China in recent 50 years. Acta Meteorol Sinica 62(2):228–236

Wang XG, Han JG, Dong YP (2005) Recent grassland policies in China: an overview. Outlook on Agric 34(2):105–110

Wang XH, Zheng D, Shen YC (2008) Land use change and its driving forces on the Tibetan Plateau during 1990–2000. CATENA 72(1):56–66

Wang J, Guo N, Cai DH, Deng ZY (2009) The effect evaluation of the program of restoring grazing to grasslands in Maqu County. Acta Ecol Sin 29(3):1276–1284

Weltzin JF, Loik ME, Schwinning S, Williams DG, Fay PA, Haddad BM (2003) Assessing the response of terrestrial ecosystems to potential changes in precipitation. Bioscience 53(10):941–952

Yang XP (2010) Climate change and desertification with special reference to the cases in China. Changing Climates, Earth Systems and Society. Springer, Berlin, pp 177–187

Yang YH, Fang JY, Ma WH, Wang W (2008) Relationship between variability in aboveground net primary production and precipitation in global grasslands. Geophys Res Lett 35(23):L23710

Yu DY, Shi PJ, Han GY, Zhu WQ, Du SQ, Xun B (2011) Forest ecosystem restoration due to a national conservation plan in China. Ecol Eng 37(9):1387–1397

Zha Y (2004) Assessment of grassland degradation near Lake Qinghai, West China, using Landsat TM and reflectance spectra data. Int J Remote Sens 25(20):4177–4189

Zhai PM, Zhang XB, Wan H, Pan XH (2005) Trends in total precipitation and frequency of daily precipitation extremes over China. J Clim 18(7):1096–1108

Zhang GL, Dong JW, Xiao XM, Hu ZM, Sheldon S (2012) Effectiveness of ecological restoration projects in Horqin Sandy Land, China based on SPOT-VGT NDVI data. Ecol Eng 38(1):20–29

Zhong HP, Fan JW, Yu GR, Han B, Hu ZM, Yue YZ (2005) Progress of carbon cycle research in grassland ecosystem. Plumbing Connection 13(z1):37–73

Zhou CY, Zhang DQ, Wang SY, Zhou GY, Liu SZ, Tang XL (2004) Diurnal variations of fluxes of the greenhouse gases from a coniferous and broad-leaved mixed forset soil in Dinghushan. Acta Ecol Sinica 24(8):1738–1741

Zhou G, Wang Y, Wang S (2010) Responses of grassland ecosystems to precipitation and land use along the Northeast China Transect. J Veg Sci 13(3):361–368

Zhu ZD (1997) Global change and desertification. Earth Sci Front 4(1):213–219

Zhu W, Pan Y, He H, Yu D, Hu H (2006) Simulation of maximum light use efficiency for some typical vegetation types in China. Chin Sci Bull 51(4):457–463

Chapter 6
Grassland Degradation Remote Sensing Monitoring and Driving Factors Quantitative Assessment in China from 1982 to 2010

Abstract Remote sensing monitoring of grassland degradation will indicate the grassland degradation status of China. At the same time, quantitative assessment of the driving factors will benefit to the understanding of degradation mechanism and grassland degradation control. In this study, net primary productivity (NPP) and grass coverage were selected as indicators to analyze grassland degradation dynamics. And we designed a method to assess the driving force of grassland degradation based on NPP. Specifically, the potential NPP and LNPP (NPP loss because of human activities), which is the difference between potential NPP and actual NPP, were used to calculate the contribution of climate and human factors to grassland degradation, respectively. Results showed that grassland degradation area accounted for 22.7% of the total grassland area in China from 1982 to 2010. The contribution of climate change and human activities to grassland degradation was almost equilibrium (47.9% vs. 46.4%). Overall, on the grassland restoration, human activities were the dominant driving factors, accounting for 78.1%, whereas the contribution of climate change was only 21.1%. However, there are obviously spatial heterogeneous on driving factors. And the contribution of climate change was larger than human activities. But for the grassland restoration, human activities were the dominant factors. Warm–dry climate was harmful to grass growth but useful restoration measurements were benefit to grassland restoration. Methods in this study can be widely used in other regions of grassland degradation evaluation.

Keywords Degraded grassland · Driving mechanism · Climate change · Human activities · Carbon sequestration

6.1 Introduction

Grasslands, one of the most common types of vegetation in the world, account for nearly 20% of the global land surface (Scurlock and Hall 1998). Human food production and to a lesser extent climate change have profoundly influenced grasslands (Conant et al. 2001). China has 3.93 million km^2 of grasslands, which account for about 40% of China's total land area. However, approximately 866,700 km^2 of China's grassland is degraded (Bao et al. 1998; Scurlock and Hall 1998). Recent

© Springer Nature Singapore Pte Ltd. 2020
W. Zhou et al., *Remote Sensing Monitoring and Evaluation of Degraded Grassland in China*, Springer Geography,
https://doi.org/10.1007/978-981-32-9382-3_6

studies have shown that nearly 90% of the grasslands in northern China are degraded to some extent (Nan 2005). Grassland degradation is mostly attributed to overgrazing and conversion of grassland to cropland as well as unregulated collection of fuel and medicinal plants (Akiyama and Kawamura 2007). Furthermore, drought, locust attacks, and rodent activities as well as climate change contribute to grassland degradation (Liu et al. 2004).

Grassland degradation is related to relevant issues such as declining productivity, biodiversity loss, land degradation, and declining ecosystem services (Turner et al. 2001). Although the cause of grassland degradation is complex, overgrazing is regarded as the leading cause (Teague and Dowhower 2003). Meanwhile, the changes in vegetation and soil because of overgrazing are accompanied by a decrease in primary production of vegetation (Snyman and Fouché 1991). Furthermore, climate change especially water and temperature will influence the length of the growing season as well as physiological processes, primary productivity, community composition, and plant diversity which is affected (Lemmens et al. 2006; Levy et al. 2004; Saleska et al. 1999). In fact, climate drying that has occurred in recent decades in north China adds further stress to the ecosystem (Chen and Tang 2005).

To alleviate the multifaceted environmental degradation, the Chinese government has launched several ecological restoration programs. Especially the Grain to Green Program (GTGP) and Returning Grazing Land to Grassland Program (RGGP) have obtained considerable achievement and deeply affected the structure and function of grassland ecosystem (Liu et al. 2008; Wang et al. 2011a, b). GTGP aims to convert cropland to forest and grassland in fragile areas, initiated since 1999 (Ferraro and Kiss 2002), whereas RGGP, initiated since 2003, aims to alleviate grazing pressure in degraded grassland through forbidding grazing and rotational grazing (Tong et al. 2004). Long periods of forbidden grazing are expected to increase plant coverage (Wang et al. 2011a, b), and previously degraded grasslands in Inner Mongolia (IM) have been restored to their 1960s' level after three years of protection from grazing (Jiang et al. 2006).

Grassland degradation monitoring is traditionally studied by field investigation, through which contributing factors are identified (Li 1997). This method is inefficient and costly because grassland usually covers a large spatial region (Asrar et al. 1986), and the results are unreliable. In contrast, remote sensing monitoring is much more efficient in assessing grassland degradation (Alfredo et al. 2002). However, the contribution of the two factors on grassland degradation is unclear at present. Therefore, a method to assess the driving contribution is crucial to monitor grassland degradation.

Recent studies have analyzed human-induced vegetation degradation based on rainfall use efficiency (RUE) method (Prince et al. 2004; Symeonakis and Drake 2004). However, RUE is an oversimplified empirical indicator and provides results that are not very reliable. Several studies also have used vegetation dynamics to distinguish human-induced desertification from climate change (Wessels et al. 2007; Xu et al. 2010). The vegetation dynamics are the most intuitive manifestation of land degradation. Meanwhile, NPP is sensitive to both climatic change and human activities (Schimel 1995). In this study, NPP and coverage were selected as vegetation dynamic indicators to reflect grassland degradation situation. In order to calculate the driving contribution, potential NPP and LNPP (NPP loss caused by human activities), which is the difference between potential NPP and actual NPP, are used to assess the relative roles of climate change and human activities in grassland degradation.

This study aims to make clear the spatial–temporal characteristic of grassland degradation in China and then determine the dominant factor of grassland degradation. Meanwhile, the results of this study will provide a deeper and more comprehensive knowledge of grassland degradation as well as useful suggestions provide recommendations for grassland resource management and sustainable development.

6.2 Methodology

6.2.1 Study Areas

The Global Land Cover 2000 dataset (GLC 2003) indicated that China's grassland area is 3.35 million km^2, covering approximately 35% of the country's total land area. It mainly distributed in the northwest China and Tibet plateau. The nine provinces regions in China, namely IM, Xinjiang, Qinghai, Tibet, Gansu, Shaanxi, Ningxia, Yunnan, and Sichuan, account for 94% of the total grassland in China (Fig. 6.1).

Fig. 6.1 Location of the study area and the distribution of grassland types in China

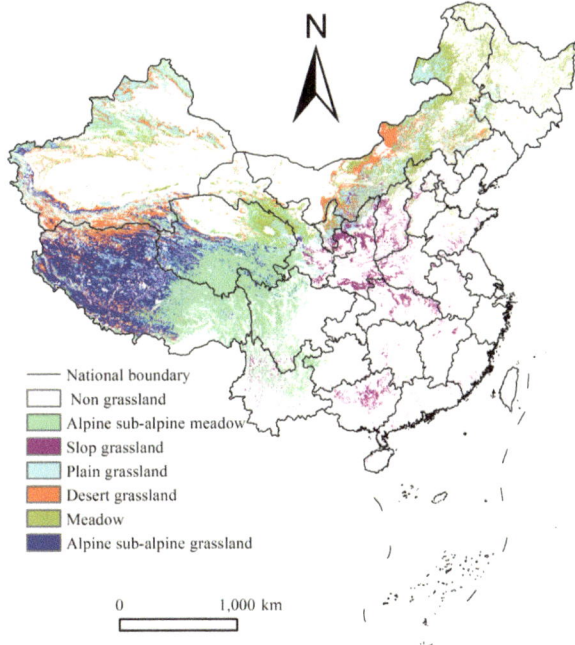

Northwest China is characterized by arid and semi-arid climate and large temperature differences between day and night. Altai, Tianshan, Kunlun, Qilian and other high-precipitation mountains have blocked the atmospheric circulation in the rain shadow, forming broad desert basins such as Tarim, Junggar and Qaidam (Shi et al. 2007). Grassland degradation is serious because of the land-use change, overgrazing, and global warming.

Tibet Plateau is the highest contiguous area of the world with approximately 1.4 million km^2 in land area perched 4500 m above sea level (Huddleston et al. 2003). It is characterized by a subtropical to temperate mountain climate unique to the Qinghai–Tibet Plateau (Chen et al. 2006). Its surface temperature is relatively low because of its high altitude. However, Tibet Plateau has been experiencing a warming trend since the mid-1950s. Precipitation in Tibet Plateau is relatively low and extremely variable in time and space. Natural vegetation in Tibet Plateau varies greatly and comprises forests, grasslands, and shrubs, which are very sensitive to environmental changes and human activities. To date, Tibet Plateau approximately has 425,100 km^2 of degraded grassland, and severely degraded grassland accounts for approximately 16% of the degraded grassland (Wang 2006).

6.2.2 Data Sources and Processing

6.2.2.1 Normalized Difference Vegetation Index (NDVI) Data and Post-processing

We used NDVI data and geospatial meteorological data as input data for the Carnegie–Ames–Stanford Approach (CASA) model to calculate the actual NPP (Potter et al. 1993). The study period was from 1982 to 2010. Two types of NDVI dataset were used in this study: moderate-resolution imaging spectroradiometer (MODIS)-NDVI (MOD13A2) data with 1 km × 1 km spatial resolution, covering the periods from 2001 to 2010 and downloaded from the earth observing system data gateway (http://edcimswww.cr.usgs.gov/pub/imswelcome/); and advanced very high-resolution radiometer global inventory modeling and mapping studies (GIMMS)-NDVI data with 8 km × 8 km resolution, covering the periods from 1982 to 2006.

A regression model for the entire pixel was established based on the two types of NDVI dataset (through the overlap time period of the two types NDVI datasets and total of 72 months from 2001 to 2006) to produce a long period of NDVI datasets from 1982 to 2010. Firstly, Savitzky–Golay filters were used to smooth the NDVI data and reduce image noises. Secondly, the nearest neighbor method was used to resolve the different spatial resolutions of the two types of NDVI dataset. Finally, maximum-value compositing was used to merge the MODIS_NDVI value from 16 and GIMMS_NDVI 15 days to produce the monthly NDVI datasets. The two NDVI products were reprojected to Albers equal-area projection based on the WGS-84 datum using ArcGIS V10.1 software (ESRI, California, USA).

6.2.2.2 Meteorological Data

Meteorological data from 1982 to 2010, including average monthly temperature and precipitation for 680 stations as well as total solar radiation data for 102 stations, were obtained from China Meteorological Data Sharing Service System. Ordinary Kriging interpolation was used to interpolate the meteorological data into grid at 1 km × 1 km spatial resolution. The driving meteorological data for the Miami memorial model to estimate the potential NPP (Lieth 1975) were annual temperature and precipitation. This can be calculated through incorporating the 12-month temperature and precipitation.

6.2.2.3 Field Survey of NPP

We sampled 51 sites across northwest China from April to August of 2009. At each site (20 m × 20 m), we set four quadrates (5 m × 5 m) and marked as S1, S2, S3, and S4. In order to calculate the NPP of all plants, we investigated twice in the quadrates, early April in quadrates S1 and S3, and later August in quadrates S2 and S4. The biomass increment for grassland was obtained by the difference between the maximum (measured in August) and the minimum biomass values (measured in April).

All plants in each quadrate were harvested to determine aboveground biomass. To determine underground biomass, nine soil cores (8 cm diameter) were used to collect samples at 10-cm intervals. The biomass samples were oven-dried at 65 °C to a constant mass and weighed to the nearest 0.1 g. At last the biomass increment of aboveground and underground was converted to carbon content using a conversion factor of 0.45. These field observation data were used to verify the accuracy of the NPP estimated by model.

6.2.3 Model

(1) Calculation of actual NPP based on CASA model

Actual NPP was calculated using the CASA model, which is a light-use efficiency model based on the resource-balance theory (Field et al. 1995). In the CASA model, NPP is the product of absorbed photosynthetically active radiation (APAR) and light-use efficiency (ε). The basic principle of the model can be described by the following formula:

$$\mathrm{NPP}(x, t) = \mathrm{APAR}(x, t) \times \varepsilon(x, t) \tag{1}$$

where x is the spatial location (pixel number), t is time, APAR is the canopy-absorbed incident solar radiation integrated over a given time period (MJ m^{-2}), and $\varepsilon(x, t)$

is the actual light-use efficiency (gC MJ^{-1}). APAR(x, t) is calculated based on the following formula:

$$APAR(x, t) = SOL(x, t) \times FPAR(x, t) \times 0.5 \tag{2}$$

where SOL(x, t) is total solar radiation (MJ m^{-2}) of pixel x in time t, 0.5 is the proportion of SOL available for vegetation (wavelength range of 0.38–0.71 μm), and FPAR(x, t) is the fraction of PAR absorbed by vegetation and calculated by the linear function of NDVI simple ratio (SR) as follows:

$$SR = [1 + NDVI(x, t)] / [1 - NDVI(x, t)] \tag{3}$$

$$FPAR(x, t) = \min\left[\frac{SR(x, t) - SR_{min}}{SR_{max} - SR_{min}}, 0.95\right] \tag{4}$$

where SR$_{min}$ is SR for bare land and is set to 1.05 for all grid cells, and SR$_{max}$ is for land independent of vegetation types, and in this study, the SR$_{max}$ for grassland is 4.46 (Potter et al. 1993; Piao et al. 2006). A cap of 0.95 was imposed on FPAR to reflect a finite upper limit to leaf area. The algorithm of $\varepsilon(x, t)$ can be expressed as follows:

$$\varepsilon(x, t) = T_{\varepsilon 1}(x, t) \times T_{\varepsilon 2}(x, t) \times W_{\varepsilon}(x, t) \times \varepsilon_{max} \tag{5}$$

where s and $T_{\varepsilon 2}(x, t)$ are temperature stress coefficients and $W_{\varepsilon}(x, t)$ is the water stress coefficient that indicates the reduction of light-use efficiency caused by moisture factor. A more detailed description of $T_{\varepsilon 1}(x, t)$, $T_{\varepsilon 2}(x, t)$, and $W_{\varepsilon}(x, t)$ can be found in Yu et al. (2011).

ε_{max} indicates the maximal light-use efficiency under ideal conditions. And ε_{max} is easily affected by actual temperature and moisture conditions and differs greatly compared with real conditions (Paruelo et al. 1997). The value of ε_{max} for grassland is 0.542 in this study in accordance with the study of Zhu et al. (2006).

(2) Validation of CASA model

Validation was conducted by comparing the field observation data for grassland with the estimated data by CASA model. Figure 6.2 presents the results of the correlation between the observed NPP and estimated NPP. The correlation was significant ($R^2 = 0.7043$, $P < 0.001$), which indicates that the model's estimation accuracy is reliable in grassland of China.

(3) Estimation of potential NPP based on the Miami model

Although several models have been developed to estimate NPP, these are based on different climate factors. The Miami model (Lieth 1975) was the first widely used model derived from least squares correlations between measured NPP data and the corresponding temperature and precipitation data. Over the past 30 years, this model

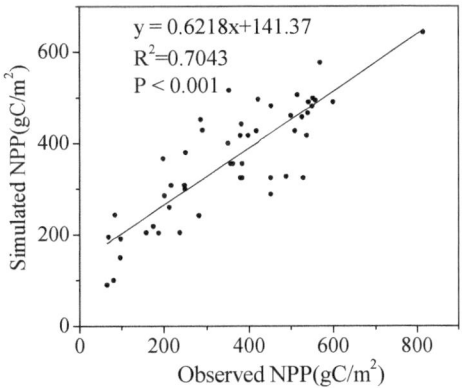

Fig. 6.2 Validation of the CASA model accuracy in grassland of China through the correlation analysis between estimated NPP based on the CASA model and the field observation NPP (gC/m^2) in July 2009

has been used as a baseline dataset and has yielded "reasonable estimates" of global patterns of productivity (Adams et al. 2004). The basic principle of the Miami model can be described by the following formula:

$$NPP_t = 3000/(1 + e^{1.315 - 0.119t}) \tag{6}$$

$$NPP_p = 3000/(1 + e^{-0.000664p}) \tag{7}$$

where t is annual mean temperature (°C) and p is annual precipitation (mm). Liebig's restrictive factor law indicates that the simulated NPP value is either NPP_t or NPP_p, whichever is smaller. Miami model was originally used as a tool for estimating the spatial distribution of NPP based on long-term climate data (Malmström et al. 1997). It has been used formally to analyze year-to-year variation in NPP response to inter-annual changes in climate (Krausmann et al. 2009).

6.2.4 Calculation of Fractional Vegetation Cover

Fractional vegetation cover (f_v), which is defined as the projected area of vegetation per unit of ground area, is an important ecological parameter to evaluate grassland degradation. According to Gutman and Ignatov (Gutman and Ignatov 1998), f_v is calculated from NDVI using a linear mixture model with two end members representing fully vegetated land surface and bare ground. The formula is illustrated as follows:

$$f_v = \frac{NDVI - NDVI_{min}}{NDVI_{max} - NDVI_{min}} \tag{8}$$

where $NDVI_{min}$ is the minimum NDVI corresponding to 0% of vegetation cover or bare land, and $NDVI_{max}$ is the maximum one 100% vegetation cover. The annual mean vegetation cover for each pixel can be calculated from this method.

6.2.5   Slope of Grassland NPP

Vegetation dynamics measured by NPP are the most intuitive manifestation of grassland degradation. Equation (9) can be used to calculate the slope of PNPP and LNPP by ordinary least square estimation (Ma and Frank 2006). The formula is expressed as follows:

$$Slope = \frac{n \times \sum_{i=1}^{n} i \times NPP_i - \left(\sum_{i=1}^{n} i\right)\left(\sum_{i=1}^{n} NPP_i\right)}{n \times \sum_{i}^{n} i^2 - \left(\sum_{i=1}^{n} i\right)^2} \tag{9}$$

where i is 1 for year 1982, 2 for year 1983, and so on; n is 29 for years 1982 to 2010; and NPP_i is the value of annual NPP in time of i year.

To determine whether the NPP increased or decreased during the study period, we estimated the change in NPP for each pixel using the following equation:

$$\Delta NPP = (n - 1) \times slope \tag{10}$$

where $n = 29$ years, representing the study period from 1982 to 2010.

6.2.6   Grassland Degradation Status Evaluation

Grassland degradation leads to the ecosystem species decrease, community structure simplification, plant height, and coverage and productivity decline, as well as species diversity decrease. In this study, grassland coverage and NPP changes were selected as grassland degradation monitoring indicators, according to national standards of Peoples' Republic of China "parameters for degradation, sandification and salification of rangelands." Based on the above standard, grassland degradation was classified to four classes—non-degradation, slight degradation, moderate degradation, and severity degradation; grassland restoration also was classified as four classes—non-restoration, slight restoration, moderate restoration, and severity restoration.

6.2.7 Scenarios Design and Quantitative Assessment Method of Grassland Degradation

Quantitative assessment of the relative roles of two factors in grassland degradation can be possible when the relation between the change in NPP induced by climate change and human activities is identified. This quantitative assessment approach was built based on the method by Xu et al. (2010).

In this study, we defined three kinds of NPP. The first is actual NPP, which is calculated using the CASA model and is used to evaluate grassland dynamics based on the slope of the actual NPP (S_A). The second is potential NPP, which represents the change in NPP caused by climate change, and is estimated using the Miami model. The third is the loss of NPP because of human activities (LNPP), which equals the difference between potential and actual NPP (i.e., $LNPP = NPP_{Miami} - NPP_{CASA}$) and reflects the effects of human activities on vegetation productivity.

The effect climate changes on grassland degradation can be gauged using the slope of potential NPP (S_P). A positive S_P indicates that the inter-annual climate change during this period is beneficial to grass growth, whereas a negative S_P demonstrates that the inter-annual climate change reduces grass growth.

If S_L (slope of loss NPP) is negative, that is, the loss of NPP induced by human activities decreased during the study period; then, human activities are beneficial to grass growth, which indicates that the grassland restoration was human dominated. Conversely, a positive S_L indicates that human activities are harmful to grass growth, and human-dominated grassland degradation occurs. Finally, eight scenarios were created based on S_P and S_L in this study (Table 6.1).

6.3 Implementations and Discussions

6.3.1 Spatial Distribution of Grassland Degradation Status

The spatial distribution of grassland degradation from 1982 to 2010 in China was shown in Fig. 6.3. The severity degradation mainly distributed in the regions of Kunlun Mountains, the northwest and central of Tibet, the Tianshan Mountains, the desert grassland in the west of IM, and the Hexi Corridor region. Statistics showed that the severity degradation grassland accounted for 7.1% of the total grasslands area of China. The moderate degradation, accounting for 5.6%, located in the northwest of Tibet Plateau, central region of IM, Hulun Buir of IM. Slight degradation accounted for 10.0%, including west of Tibet Plateau, northwest of Xinjiang, central region of IM, and northeast of China.

Regions of grassland occurred obvious restoration mainly distributed in the west of Hulun Buir, Tianshan Mountains, west of Qinghai, west of Xinjiang, and south of Tibet, and the above regions accounted for 6.5% of total grassland area. Moderate

Table 6.1 Methods for assessing the driving factors of grassland restoration and/or degradation in eight scenarios

		S_P	S_L	Relative role of climate change (%)	Relative role of human activities (%)
Grassland restoration	Scenario 1	>0	>0	100	0
	Scenario 2	<0	<0	0	100
	Scenario 3	>0	<0	$\frac{\Delta \text{NPP}_P}{\Delta \text{NPP}_P + \Delta \text{NPP}_L} \times 100\%$	$\frac{\Delta \text{NPP}_L}{\Delta \text{NPP}_P + \Delta \text{NPP}_L} \times 100\%$
	Scenario 4	<0	>0	Error	Error
Grassland degradation	Scenario 1	<0	<0	100	0
	Scenario 2	>0	>0	0	100
	Scenario 3	<0	>0	$\frac{\Delta \text{NPP}_P}{\Delta \text{NPP}_P + \Delta \text{NPP}_L} \times 100\%$	$\frac{\Delta \text{NPP}_L}{\Delta \text{NPP}_P + \Delta \text{NPP}_L} \times 100\%$
	Scenario 4	>0	<0	Error	Error

Note S_P is the change trend of potential NPP (i.e., potential NPP was only influenced by climate factors), and S_L is the change trend of the loss of NPP, which equals the difference between potential and actual NPP (i.e., LNPP = $\text{NPP}_{\text{Miami}}$ − NPP_{CASA}). ΔNPP_P is the total increase or decrease of potential NPP during 1982–2010 which can be calculated using Eq. (10), and ΔNPP_L is the total increase or decrease of LNPP. Scenario 1: Climate-induced grassland restoration or degradation; Scenario 2: Human-induced grassland restoration or degradation; Scenario 3: Grassland restoration or degradation resulted from the interaction of climate change and human activities; Scenario 4: Both climate change and human activities actually benefit the grassland degradation or restoration, but in this method, restoration or degradation occurred; this judgment is wrong

Fig. 6.3 Spatial distribution of grassland degradation situation during 1982–2010

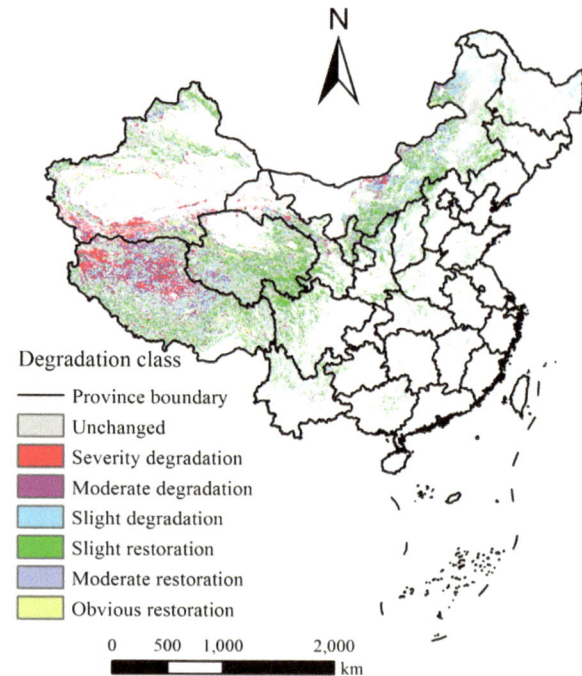

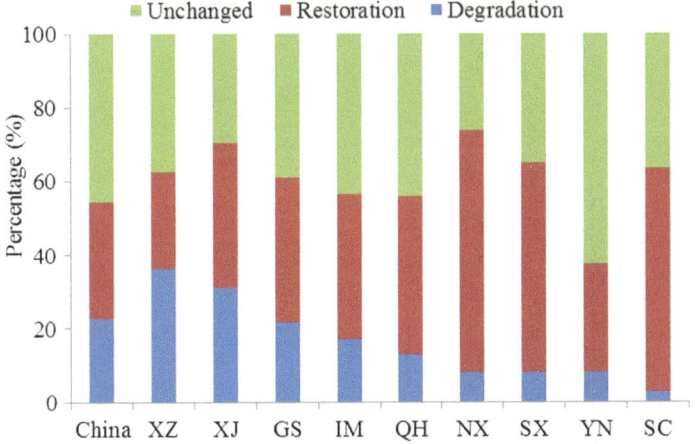

Fig. 6.4 Grassland degradation situation in the main animal husbandry provinces of China

restoration regions distributed in the central region of Tibet, accounting for 9.8%. Slight restoration grasslands were widespread, accounting for 19.8%.

Statistical analysis showed that there were 22.7% of grassland occurred degradation, restoration grassland area was 31.7%, and the rest of 45.6% remained unchanged. Moreover, the grassland degradation and restoration area were also statistical analyzed in the nine provinces. Grassland degradation area was largest in Tibet, accounting for 36.4% of the total grasslands area of Tibet, while restoration area accounting for 26.1%, and the rest of 37.5% remained unchanged. And for the rest of eight provinces, Xinjiang, Gansu, IM, Qinghai, Ningxia, Shaanxi, Yunnan, and Sichuan, the degradation areas accounted for 32.1%, 21.8%, 17.1%, 13.1%, 8.22%, 8.1%, and 2.7%, respectively (Fig. 6.4).

6.3.2 Quantitative Assessment of Grassland Degradation or Restoration in China from 1982 to 2010

The relative roles of climate change and human activities in grassland degradation or restoration were assessed based on the methods listed in Table 6.1.

The driving contribution of climate change to grassland degradation was shown in Fig. 6.5a. On the spatial distribution, climate-dominated grassland degradation distributed in the western region of Tibet, central and west of IM, Hexi Corridor. Overall, climate-dominated degradation accounted for 47.9% of the total degradation area. Human-dominated degradation mainly distributed in the northern of Tibet, north of Xinjiang, and east of Hulun Buir of IM (Fig. 6.5b). Moreover, 46.3% of grassland degradation was derived from human activities. And the rest of 11.8% was attributed to the error.

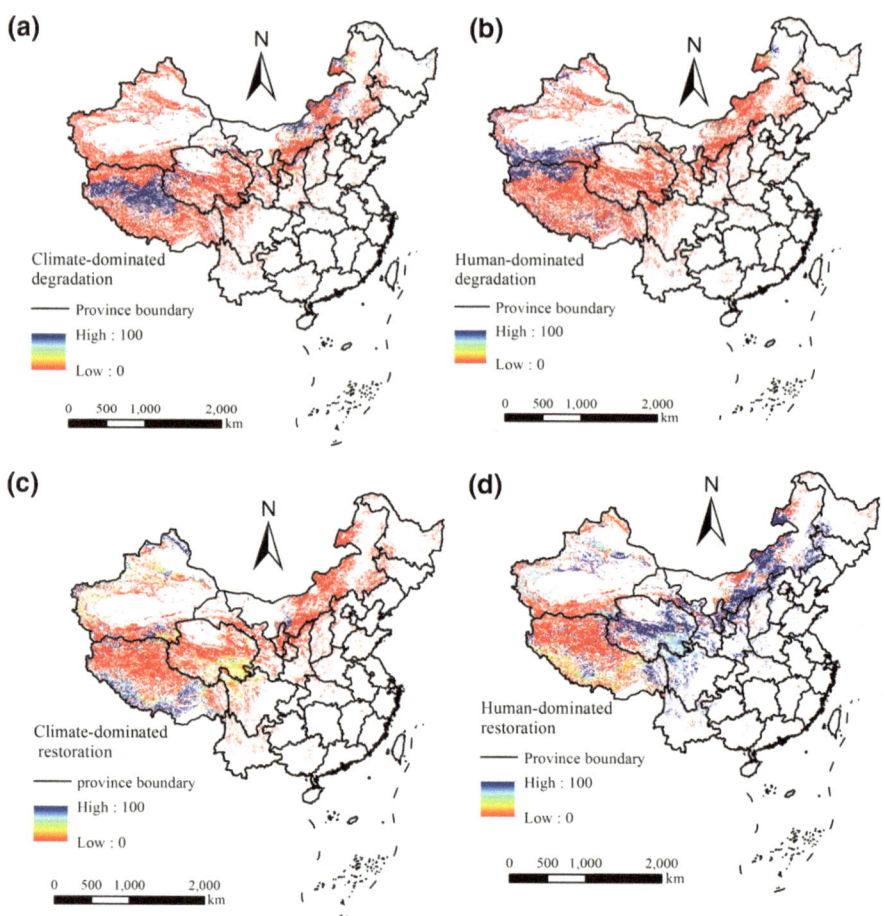

Fig. 6.5 Spatial distribution of the contributions of **a** climate and **b** human factors in grassland degradation, and the contributions of **c** climate and **d** human factors in grassland restoration

For the grassland restoration, the contribution of climate change is shown in Fig. 6.5c. And on the spatial distribution, climate-dominated restoration mainly included west and south of Tibet, Altai Mountains, Kunlun Mountains, northwest of Mu Us desert lands. Statistical analysis indicated that 21.1% of restoration was attributed to climate change. However, human-dominated restoration was widespread in the most regions of IM, Ningxia, Shaanxi, Qinghai, South of Gansu, Sichuan, Yunnan, southeast of Tibet, Tianshan Mountains (Fig. 6.5d). And human-dominated restoration accounted for 78.1%; and the error derived from the quantitative assessment method was 0.8%.

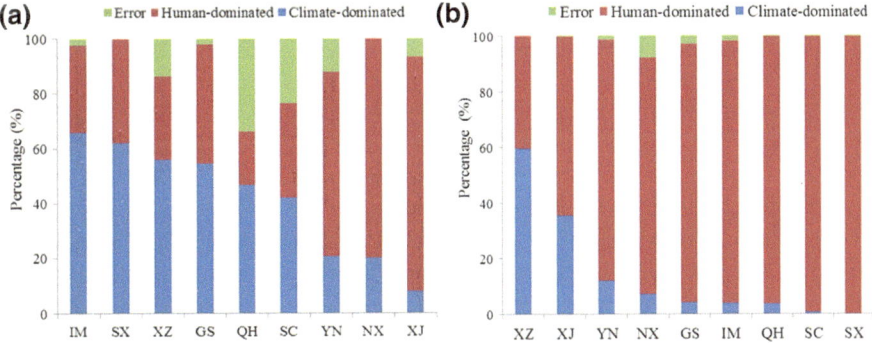

Fig. 6.6 Driving factors of **a** grassland degradation and **b** restoration in the nine animal husbandry provinces

6.3.3 Comparative Analysis of the Driving Contribution of Two Factors in the Nine Provinces

Grasslands in the nine provinces accounted for 94% of the total grasslands in China. The relative roles of climate and human factors in grassland degradation or restoration varied greatly in the nine provinces. For the grassland degradation, provinces of the contribution of climate change were larger than human activities included IM, Shaanxi, Tibet, Gansu, Qinghai, and Sichuan; however, provinces of human-dominated degradation included Yunnan, Ningxia, and Xinjiang (Fig. 6.6a), and especially in Xinjiang, where 85.1% of degradation was attributed to human activities. For the grassland restoration, climate-dominated restoration solely occurred in Tibet (Fig. 6.6b), and its contribution reached to 59.7%, and 40.3% of restoration was attributed to human activities. While, for the rest of the eight provinces, human activities were the dominant driving factor, especially for Sichuan and Shaanxi, where more than 99% of restoration was attributed to human activities.

Overall, there are three kinds of outcomes in the nine provinces. One is climate change which was the dominant driving factor on both degradation (55.9%) and restoration (59.7%), as seen in Tibet. Another outcome, human is the dominated impact on both grassland degradation and restoration, as seen in Yunnan, Ningxia, and Xinjiang provinces. The third outcome is human-dominated restoration and climate-dominated degradation, such as in the case of in IM, Shaanxi, Gansu, Qinghai, and Sichuan.

6.3.4 Discussion

(1) **Quantitative assessment methods for grassland degradation**

The monitoring and assessment of grassland degradation traditionally mainly focused on field survey, which is inefficient. Results obtained from these traditional methods are unreliable and costly time and labor efforts, as grasslands usually cover large areas (Asrar et al. 1986). Assessing grassland degradation by remote sensing is much more efficient (Alfredo et al. 2002; Lu et al. 2007). Although previous studies have assessed grassland degradation using satellite imagery (Liu et al. 2004; Shi et al. 1999), these studies merely mapped the degradation class spatial distribution. Quantitative assessments of the grassland degradation driving mechanism at a large spatial–temporal scale are few in China.

Grassland degradation refers to the overall reduction in grassland productivity, reflected by reduction in grass coverage, grass density, and biological diversity, or by the increase in unpalatable grass species and toxic weeds, and even soil erosion (Liu et al. 2004). Therefore, in this study, NPP and coverage were selected as monitoring indicator for grassland degradation. In previous study, NPP has been used as an intuitive vegetation dynamic indicator to distinguish the impacts of climate change from human activities on land degradation (Zheng et al. 2006). Furthermore, potential NPP and LNPP were used to assess the contribution of climate change and human activities to grassland degradation.

(2) **Climate change and its impacts on grassland vegetation dynamics**

Previous studies have showed that land degradation is caused by the feedback inter-action process between climate change and human activities. In this process, climate change is an intrinsic controlling driving force, whereas human activities are seen as an external driving force that can intensify or alleviate grassland degradation to some extent (Wang 2006; Zheng et al. 2006). Climate change affects grassland vegetation dynamics mainly through variation in temperature and precipitation. Meanwhile, recent studies suggest that climate in northwest China has changed from warm–dry to warm–wet since the late 1980s (Shi et al. 2007). Zhai et al. analyzed the change in precipitation in China from 1950 to 2008 and found that precipitation increased significantly in the Tianshan Mountains, southeastern Tibet, and the western Tarim Basin (Zhai et al. 2005). Qi et al. also concluded that the climate in Xinjiang exhibited a wet trend from 1940 to 2008 (Qi et al. 2012). Our findings confirm the earlier studies by showing evidence of the increase in annual precipitation from 1982 to 2010 in northwest China, except in the south of Tibet, Gansu, and Ningxia provinces; however, rising temperatures occurred in most areas of China (Fig. 6.7). At the same time, our findings show that inter-annual climate change benefits grassland restoration, contributing up to 21.1%, and mainly distributed in the south of Tibet Plateau and north of Xinjiang province, as shown in Fig. 6.6c, which indicated the warm–wet climate in northwest China is beneficial to vegetation growth and also be found in other studies (Zheng et al. 2006). However, the decline in precipitation was obvious in east and south China, especially in IM, Shaanxi, Sichuan, and Yunnan. Simultaneously, climate-dominated grassland degradation was also significant (Fig. 6.5a). Additionally, 44.9% of grassland degradation was caused by the rising of temperature and decline of precipitation, especially in central and west of Tibet, Central of IM (Fig. 6.5a). Therefore, precipitation is a factor that limits vegetation growth especially in dry land (Herrmann et al. 2005).

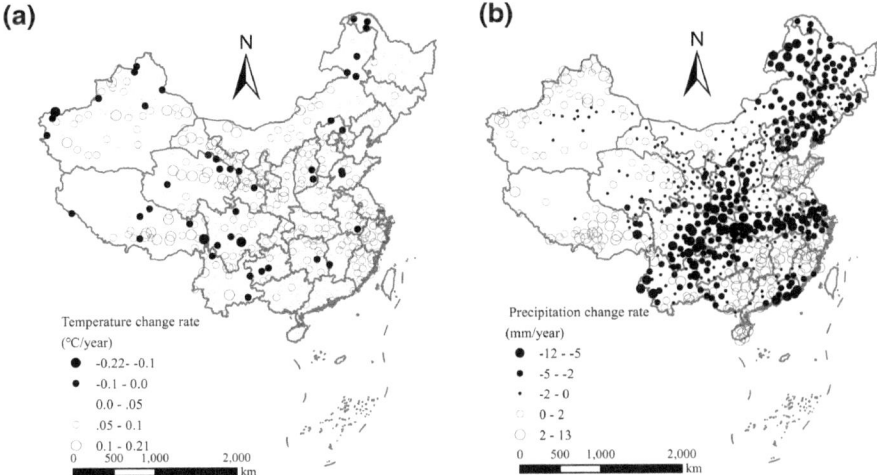

Fig. 6.7 Changing tendency of temperature **a** and precipitation **b** in China during 1982–2010

(3) **Impacts of grazing and restoration projects on grassland degradation or restoration**

Human activities are a pervasive catalyst for land degradation, as land-use cover change and management measures largely govern the sustainability of a given land (Foley et al. 2005). Particularly in recent decades, a series of ecological restoration projects were implemented to control grassland degradation. Two of them, the GTGP and RGGP, aimed at alleviating grazing pressure in degraded grasslands, mainly through fenced enclosure and rotational grazing to allow natural reestablishment of the native vegetation. Restoration programs have positive effects, as reported in many previous studies. For example, in a degraded Alxa desert steppe with grazing exclusion for four years, hay yield increased by 134.7%, and vegetation coverage increased by 97% after fencing (Ta et al. 2008). Wang et al. (2009) found that grass is clearly restored in Maqu County after RGGP implementation. In our study, human activities were the dominant driving factor to grassland restoration from 1982 to 2010, contributing up to 78.1%, especially in Sichuan, Shaanxi, Qinghai, IM, and Gansu, where human-dominated grassland restoration was larger than 93%.

Undoubtedly, these positive measures have promoted grassland restoration. However, the difference of restoration projects and human positive management also leads to the difference in implementation effects. According to the statistical yearbook, the artificial grass, forbidden grazing, and fencing area in IM were largest, contributing up to 40%, 38%, and 40% of the total corresponding area of China. Furthermore, the forbidden grazing area in Shaanxi and Gansu accounted for 12.5% and 10% of total forbidden grazing area in China. Meanwhile, the quantitative assessment of grassland restoration showed that the contribution of human activities reached up to 93% in the above provinces. However, the implementation efforts of restoration in Tibet were relative low, and the corresponding contribution was merely 40.3% to grassland restoration. However, grazing pressure has increased with the population growth, and overgrazing has ultimately led to new degradation in some regions. Our findings indicated that over 46.4% of grassland degradation was caused by human activities, mostly in regions with high grassland coverage.

In summary, our model suggests that human activities dominated grassland restoration accounting for 78.1%; for the grassland degradation, the contribution of climate and human factors was nearly equilibrium (47.9% vs. 46.4%). And the driving contribution was difference for the nine provinces. Therefore, China should adopt further steps toward sustainable grassland development by enhancing the effectiveness of ecological restoration programs in a large area, except in severely degraded local regions. Measures, such as establishing artificial grassland, cultivating high-yielding grasses, and controlling locust and mice, should be implemented to protect grasslands.

6.4 Conclusions

In this paper, grassland NPP and coverage were used to evaluate grassland dynamics; potential NPP and LNPP were applied to measure the contributions of climate change and human activities to grassland degradation or restoration. From 1982 to 2010, 22.7% of the total grassland was degraded; restoration area reached to 31.7%. Human activities dominated grassland restoration, its contribution reached up to 78.1%, while the contribution of climate change was merely 21.1%. For grassland degradation, the contribution of climate and human factors was nearly equilibrium (47.9% vs. 46.4%).

The relative roles varied greatly in the nine provinces that comprise 94% of the total grasslands area in China. On the grassland degradation, the effect of climate change was larger than human activities mainly including the following provinces, IM, Shaanxi, Tibet, Gansu, Qinghai, and Sichuan; human-dominated grassland degradation distributed in Yunnan, Ningxia, and Xinjiang, especially Xinjiang, 85.1% of degradation was attributed to human activities. On the grassland restoration, climate change was the dominant factor solely in Tibet and the contribution was 60%; for the rest of provinces, human activities were dominated factor. In sum, the methods built in this study can be used in other regions, and the effectiveness of ecological restoration projects should be enhanced not only to control grassland degradation, but also to increase carbon sequestration.

References

Adams B, White A, Lenton TM (2004) An analysis of some diverse approaches to modelling terrestrial net primary productivity. Ecol Model 177(3):353–391

Akiyama T, Kawamura K (2007) Grassland degradation in China: methods of monitoring, management and restoration. Grassl Sci 53(1):1–17

Alfredo CD, Emilio CW, Ana C (2002) Satellite remote sensing analysis to monitor desertification processes in the crop-rangeland boundary of Argentina. J Arid Environ 52(1):121–133

Asrar G, Weiser RL, Johnson DE, Kanemasu ET, Killeen JM (1986) Distinguishing among tall-grass prairie cover types from measurements of multispectral reflectance. Remote Sens Environ 19(2):159–169

Bao WZ, Shan W, Yang XD, Sun H, Lan YF (1998) Ecological crises facing the grassland resources in northern China and their solutions. Grassl China (2):68–71

Chen Y, Tang H (2005) Desertification in north China: background, anthropogenic impacts and failures in combating it. Land Degrad Dev 16(4):367–376

Chen S, Liu Y, Thomas A (2006) Climatic change on the Tibetan Plateau: potential evapotranspiration trends from 1961–2000. Clim Change 76(3–4):291–319

Conant RT, Paustian K, Elliott ET (2001) Grassland management and conversion into grassland: effects on soil carbon. Ecol Appl 11(2):343–355

Ferraro PJ, Kiss A (2002) Direct payments to conserve biodiversity. Science 298(5599):1718–1719

Field CB, Randerson JT, Malmström CM (1995) Global net primary production: combining ecology and remote sensing. Remote Sens Environ 51(1):74–88

Foley JA, DeFries R, Asner GP, Barford C, Bonan G, Carpenter SR (2005) Global consequences of land use. Science 309(5734):570–574

GLC (2003) Global landcover classification for the year 2000. http://www-gem.jrc.it/glc2000/

Gutman G, Ignatov A (1998) The derivation of the green vegetation fraction from NOAA/AVHRR data for use in numerical weather prediction models. Int J Remote Sens 19(8):1533–1543

Herrmann SM, Anyamba A, Tucker CJ (2005) Recent trends in vegetation dynamics in the African Sahel and their relationship to climate. Global Environ Change 15(4):394–404

Huddleston B, de Salvo EAP, Zanetti M, Bloise M, Bel J, Franceschini G (2003) Towards a GIS-based analysis of mountain environments and populations. Environment and Natural Resources working paper No. 10. Food and Agricultural Organisation, Rome

Jiang GM, Han XG, Wu JG (2006) Restoration and management of the Inner Mongolia grassland require a sustainable strategy. AMBIO J Hum Environ 35(5): 269–270

Krausmann F, Haberl H, Erb K, Wiesinger M, Gaube V, Gingrich S (2009) What determines geographical patterns of the global human appropriation of net primary production. J Land Use Sci 4(1–2):15–33

Lemmens C, Boeck HD, Gielen B, Bossuyt H, Malchair S, Carnol M (2006) End-of-season effects of elevated temperature on ecophysiological processes of grassland species at different species richness levels. Environ Exp Bot 56(3):245–254

Levy PE, Cannell M, Friend AD (2004) Modelling the impact of future changes in climate, CO_2 concentration and land use on natural ecosystems and the terrestrial carbon sink. Glob Environ Change 14(1):21–30

Li B (1997) The degradation of grassland in North China and its countermeasure. Sci Agric Sin 30:1–10

Lieth H (1975) Modeling the primary production of the world. In: Lieth H, Whittaker RH (eds) Primary productivity of the biosphere. Springer, Berlin, Germany, pp 237–263

Liu SL, Wang T (2007) Aeolian desertification from the mid-1970s to 2005 in Otindag Sandy Land, Northern China. Environ Geol 51(6):1057–1064

Liu Y, Zha Y, Gao J, Ni S (2004) Assessment of grassland degradation near Lake Qinghai, West China, using Landsat TM and in situ reflectance spectra data. Int J Remote Sens 25(20):4177–4189

Liu JG, Li SX, Ouyang ZY, Tam C, Chen XD (2008) Ecological and socioeconomic effects of China's policies for ecosystem services. Proc Natl Acad

Lu D, Batistella M, Mausel P, Moran E (2007) Mapping and monitoring land degradation risks in the Western Brazilian Amazon using multitemporal Landsat TM/ETM + images. Land Degrad Develop 18(1):41–54

Ma MG, Frank V (2006) Interannual variability of vegetation cover in the Chinese Heihe River Basin and its relation to meteorological parameters. Int J Remote Sens 27(16):3473–3486

Malmström CM, Thompson MV, Juday GP, Los SO, Randerson JT, Field CB (1997) Interannual variation in global-scale net primary production: testing model estimates. Glob Biogeochem Cycles 11(3):367–392

Nan ZB (2005) The grassland farming system and sustainable agricultural development in China. Grassl Sci 51(1):15–19

Paruelo JM, Epstein HE, Lauenroth WK, Burke IC (1997) ANPP estimates from NDVI for the central grassland region of the United States. Ecology 78(3):953–958

Piao SL, Mohammat A, Fang JY, Cai Q, Feng JM (2006) NDVI-based increase in growth of temperate grasslands and its responses to climate changes in China. Global Environ Change 16(4):340–348

Potter CS, Randerson JT, Field CB, Matson PA, Vitousek PM, Mooney HA (1993) Terrestrial ecosystem production: a process model based on global satellite and surface data. Global Biogeochem Cycles 7(4):811–841

Prince SD, Colstoun D, Brown E, Kravitz LL (2004) Evidence from rain-use efficiencies does not indicate extensive Sahelian desertification. Global Change Biol 4(4):359–374

Qi JG, Chen JQ, Wan SQ, Ai LK (2012) Understanding the coupled natural and human systems in dryland East Asia. Environ Res Lett 7(1):15202

Saleska SR, Harte J, Torn MS (1999) The effect of experimental ecosystem warming on CO_2 fluxes in a montane meadow. Global Change Biol 5(2):125–141

Schimel DS (1995) Terrestrial biogeochemical cycles: global estimates with remote sensing. Remote Sens Environ 51(1):49–56

Scurlock J, Hall DO (1998) The global carbon sink: a grassland perspective. Global Change Biol 4(2):229–233

Shi D, Qiao A, Sai W, Hon X, Hodgson N (1999) Applied research on use of remote sensing to study alpine grassland resource and degradation. Grassl Qinghai 8:1–6

Shi YF, Shen YP, Kang E, Li DL, Ding YJ, Zhang G (2007) Recent and future climate change in northwest China. Clim Change 80(3):379–393

Snyman HA, Fouché HJ (1991) Production and water-use efficiency of semi-arid grasslands of South Africa as affected by veld condition and rainfall. Water SA 17(4):263–268

Symeonakis E, Drake N (2004) Monitoring desertification and land degradation over sub-Saharan Africa. Int J Remote Sens 25(3):573–592

Ta L, Li J, Zhang JW (2008) Monitoring on the effect of returning grazing desert to grassland in Alashan. Pratacyltural Sci 25(2):124–127

Teague WR, Dowhower SL (2003) Patch dynamics under rotational and continuous grazing management in large, heterogeneous paddocks. J Arid Environ 53(2):211–229

Tong C, Wu J, Yong S, Yang J, Yong W (2004) A landscape-scale assessment of steppe degradation in the Xilin River Basin, Inner Mongolia, China. J Arid Environ 59(1):133–149

Turner MG, Gardner RH, O'Neill RV (2001) Landscape ecology in theory and practice: pattern and process. Springer, Berlin

Wang C (2006) Simulation on the carbon and water vapor flux of the typical ecosystem by the BIOME-BGC model. Nanjing Agricultural University

Wang J, Guo N, Cai DH, Deng ZY (2009) The effect evaluation of the program of restoring grazing to grasslands in Maqu County. Acta Ecol Sin 29(3):1276–1284

Wang S, Wilkes A, Zhang Z, Chang X, Lang R, Wang Y (2011a) Management and land use change effects on soil carbon in northern China's grasslands: a synthesis. Agric Ecosyst Environ 142(3–4):329–340

Wang X, Zang S, Na X (2011) Analyzing dynamic process of land use change in Ha-Da-Qi industrial corridor of China. Procedia Environ Sci 11(Part B):1008–1015

Wessels KJ, Prince SD, Malherbe J, Small J, Frost PE, VanZyl D (2007) Can human-induced land degradation be distinguished from the effects of rainfall variability? A case study in South Africa. J Arid Environ 68(2):271–297

Xu DY, Kang XW, Zhuang DF, Pan JJ (2010) Multi-scale quantitative assessment of the relative roles of climate change and human activities in desertification—a case study of the Ordos Plateau, China. J Arid Environ 74(4):498–507

Yu DY, Shi PJ, Han GY, Zhu WQ, Du SQ, Xun B (2011) Forest ecosystem restoration due to a national conservation plan in China. Ecol Eng 37(9):1387–1397

Zhai PM, Zhang XB, Wan H, Pan XH (2005) Trends in total precipitation and frequency of daily precipitation extremes over China. J Clim 18(7):1096–1108

Zheng YR, Xie ZX, Robert C, Jiang LH, Shimizu H (2006) Did climate drive ecosystem change and induce desertification in Otindag sandy land, China over the past 40 years. J Arid Environ 64(3):523–541

Zhu WQ, Pan YZ, He H, Yu DY, Hu HB (2006) Simulation of maximum light use efficiency for some typical vegetation types in China. Chin Sci Bull 51(4):457–463

Chapter 7
Grassland Degradation Restoration and Constructing Green Ecological Protective Screen

Abstract The construction of green ecological barrier is of great significance to the restoration of degraded grassland and the protection of grassland resources. The report of the Nineteenth National Congress of the Communist Party of China pointed out that we should strengthen the protection of ecosystems, implement major projects for the protection and restoration of important ecosystems, promote the comprehensive management of desertification, rocky desertification, and soil erosion, and strengthen the protection and restoration of wetlands. Degraded grassland restoration technology is huge and complex system engineering, and strong theoretical and technical guidance is needed in grassland science, ecology, restoration ecology, soil science, and other disciplines. At present, the techniques of black soil slope restoration, grassland desertification control, grazing prohibition, grazing rotation, and planting artificial grassland have been widely developed. Meanwhile, strengthening the construction of laws and regulations, clarifying the property right of grassland, and eliminating institutional overuse of grassland institutional, perfecting compensation mechanism for grassland ecological construction, adjustment of ecological restoration project based on the monitoring results of remote sensing. Ultimately, a comprehensive system of grassland industrialization and livestock products according to local conditions needs to be established.

Keywords Green ecological protective screen · Ecological compensation · Sustainable development of animal husbandry economy · Degraded grassland

7.1 Overview of the Green Ecological Protective Screen Policy in China

The Fifth Plenary Session of the Eighteenth Central Committee of the Communist Party of China puts forward the concept of green development for the first time, puts forward to abandon the extensive growth model featuring high investment, high consumption, high pollution, and low output, and consciously promotes green development, low-carbon development, and circular development, adhere to the road of sustainable development, firmly establishing "Lucid waters and lush mountains are invaluable assets." That is to improve the production capacity of ecological prod-

© Springer Nature Singapore Pte Ltd. 2020

W. Zhou et al., *Remote Sensing Monitoring and Evaluation of Degraded Grassland in China*, Springer Geography,
https://doi.org/10.1007/978-981-32-9382-3_7

ucts, transform the advantages of ecological environment into economic advantages, obtain new economic growth points from natural environmental resources, develop circular economy, and promote economic growth and social development by giving priority to environmental protection. The report of the Nineteenth National Congress of the CPC pointed out that we should strengthen ecological system protection, implement important ecological system protection and restoration projects, carry out land greening activities, promote comprehensive control of desertification, rocky desertification and soil erosion, and strengthen wetland protection and restoration. We should improve the natural forest protection system and expand the conversion of cropland to forest and grassland.

Ecological environment protection is the most important foundation for sustainable development. We must respect nature, conform to nature, protect nature, build a national ecological security screen, and realize the unity of economic, social, and ecological benefits. Ecological barrier refers to the structure and function of the ecosystem, with filter function, buffer function, partition function, water conservation function, and shelter function, which can play a role in maintaining ecological security (Li 1997; Wen et al. 2007). Therefore, we must adhere to the basic national policy of saving resources and protecting the environment, treat the ecological environment as life does, make overall plans for the management of landscape, forests, fields, lakes, and grasses, implement the most stringent ecological environment protection system, form a green development mode and lifestyle, and firmly follow the civilized development of production, prosperity, and good ecology, build a beautiful China, create a good production and living environment for the human, and contribute to global ecological security.

7.2 Restoration Technology of Degraded Grassland

7.2.1 Grassland Degradation in China

Grassland degradation is a process in which the composition, structure, and function of grassland change obviously under artificial overload such as grazing and reclamation. The original scale of energy flow is reduced, and the material circulation is out of balance, which leads to the decline of vegetation productivity, the replacement of biological composition, the degradation of soil water and soil circulation system, and the change in near-surface microclimate deterioration. Grassland degradation is a universal problem in the world (Chen et al. 2002). At the same time, grassland degradation has also become a severe ecological environment problem facing the international community (Li 1997; Wen et al. 2007). It is estimated that about 20% of grassland biomass has declined globally. However, for Chinese grasslands, due to their complex terrain, fragile ecological environment, and sensitivity to climate change, they have become one of the most degraded areas in the world. Grassland degradation in China began in the 1960s. In the middle of 1970s, the total area of degraded grassland was 15%. By the middle of 1980s, the degraded grassland area was 866 thousand and 700 km^2, accounting for 30%. In the middle of the 1990s, it reached 1 million 333 thousand and 300 km^2, reaching over 50%. Grassland ecological environment presents an increasingly deteriorating trend. It has been reported that 90% of natural grasslands in China have been degraded to varying degrees, of which moderate and severe degraded grasslands cover an area of 1.3 million km^2, and the rate of degradation is 6700 km^2/a (Ren et al. 2007). The aboveground biomass of grassland vegetation decreased from 2.2–3.0 t/ha in the 1950s to 0.7–0.9 t/ha in the 1990s (Akiyama and Kawamura 2007).

In recent years, the grassland in northwest China has shown a trend of large-scale degradation due to the increasing intensity of human activities and global climate change (Ren et al. 2007; Yan et al. 2010). Rats are rampant in grassland, black soil beaches appear in large-scale secondary bare land, water conservation capacity and grassland productivity are declining sharply, and biodiversity is declining. Shrinkage accelerated the weakening of the ecological barrier function in pastoral areas (Liu et al. 2006; Ren et al. 2007). This not only seriously restricts the sustainable development of animal husbandry and the steady increase of farmers' income, but also threatens the ecological safety of the region (Jia 2007).

In the past 30 years, the area of degraded grassland above the middle reaches 112 million hm, which is 30–50% lower than that of the 1950s, 70–80% more poisonous and harmful weeds, 15–25% less than grassland coverage in three-river source region. The area of degraded grassland in Inner Mongolia is about 250,000 km^2, accounting for 39.2% of the total available grassland, and the degraded rate is increasing at 830 km^2/a (Zhang 2017). In the western part of northeast China, grassland without degradation occupied less than 26%, that is most grassland is in a state of degeneration; the area of slightly, moderately, and severely degraded grassland reaches 4.635 million ha, 4.81 million ha, and 9.03 million ha, respectively (Zhang and Young 2011). The Yunnan–Guizhou Plateau and Tibetan Plateau are facing with serious water and wind erosion. The proportion of edible grass is only 36%. Compared with the 1950s, the grassland yield in Qinghai decreased by 30–80% in the 1980s, resulting in an annual reduction of 12 million tons of edible fresh grass, equivalent to a reduction of 8.2 million sheep (Huang et al. 2018).

In Tianzhu County of Gansu Province, compared with the 1950s, the yield of grassland decreased by 30.34% in 1997. Degraded grassland in Xinjiang has become a serious disaster of ecological environment in Xinjiang. 80% of grassland has been degraded to varying degrees. The yield of grassland has decreased by 30–60%, 37% of which belongs to serious degraded grassland. The yield of grassland has decreased by more than 60–80%, and the carrying capacity of grassland has decreased by 1.49 ha to raise a sheep (Gao et al. 2013). Due to the serious imbalance of grassland and livestock, the individual production of livestock declined. According to the 2000 Chinese Information Net, 200 million livestock died. Because of the grassland degradation, the production level of grassland animal husbandry has been seriously affected. Natural grassland livestock production in China is only equivalent 1/82 of New Zealand, 1/20 of the United States, and 1/10 of Australia.

7.2.2 Restoration of Degraded Grassland

The research on restoration and treatment of degraded grassland began in the middle of the nineteenth century. European countries took the lead in adopting fertilization and soil moisture regulation to improve grassland conditions. Australia, New Zealand, and the former Soviet Union also studied grassland restoration techniques through fertilization, irrigation, reseeding, plowing, cutting grass, burning, introduction of alien species, and so on. In the mid-twentieth century, grassland degradation was widespread in the world, and grassland restoration and reconstruction were incorporated into grassland ecosystem management objectives. Grassland restoration has shifted from single technical improvement to comprehensive system management, forming restoration and management techniques with biodiversity maintenance, community structure optimization, and soil and seed bank restoration as the main ideas (Zhang 2017). For a long time, China has mainly adopted fencing, grazing rotation, grazing, and other management measures, as well as plowing, irrigation, fertilization, reseeding, artificial seeding, topsoil transplantation, and wood elimination to restore and improve degraded grassland. In addition, all species should be symbiotic and mutually promoted. Replanting and fencing are also important (Sun et al. 2017).

The uniformity index of fencing restoration was small, but the species richness and diversity index were higher than those of the plant communities restored by replanting. In the early stage of fencing, species importance and richness index gradually increased, and then gradually decreased. But species diversity index and evenness index decrease and then increase. When the fence is restored for a period of time, the biodiversity will gradually decline, showing a certain degree of decline. At this time, moderate artificial interventions (such as Kaifeng grazing) need to promote the harmonious development of the community.

7.2.3 Restoration Technology of Degraded Grassland

Restoring degraded grassland is a huge and complex systematic engineering, which involves not only practical techniques such as grassland science, crop pasture cultivation, soil science and ecology, but also strong scientific theories, such as restoration ecology, grassland resources science, and grassland production system theory. The restoration of degraded grassland is of great significance to economic development and ecological environment protection. The restoration of degraded grassland is a strategic task for the long-term development of a country and a nation with comprehensive ecological protection, pastoral economic development, and multicultural development. The primary condition for the restoration of degraded grassland is to eliminate the pressure applied to grassland and to reduce the threshold of grassland ecosystem restoration. That is, grassland degradation is reversible. Generally, degraded grassland has the potential function of self-recovery when removed overuse pressure, but some restoration processes will last for a long time. From the perspective of ecosystem composition, it mainly includes the restoration of abiotic and biological systems. Inorganic environment restoration technologies include water restoration technologies (e.g., pollution control, disturbance removal, drainage, and irrigation), soil restoration technologies (e.g., grassland fertilization, soil improvement, surface soil stabilization, soil erosion control, soil exchange, and decomposition of pollutants), and air restoration technologies (e.g., soot adsorption, biological and chemical adsorption and so on). Restoration technologies of biological systems include vegetation (introduction of species, variety improvement, rapid propagation of plants, plant mix, plant planting), consumer (introduction of predators, control of pests and diseases), and decomposer (introduction and control of microorganisms) and ecological planning. In the practice of restoration and treatment of degraded grassland ecosystem, a variety of technologies may be applied to the same project. The most important thing is to investigate and analyze the actual situation of degraded grassland, make full use of various technologies, restore the structure of degraded grassland ecosystem as soon as possible, and then restore its function, realize its ecological, economic, and social benefits (Liu et al. 2002).

Black soil beach restoration, using the principle of classified management, that is, moderate and mild black soil beach degraded grassland mainly semi-artificial vegetation restoration; severe black soil beach degraded grassland mainly artificial vegetation reconstruction. The restoration of black soil slope is mainly based on the artificial establishment of vegetation, using artificial sowing method, after sowing the soil to suppress and cover non-woven cloth, surrounded by fencing to prohibit grazing.

Covered non-woven fabrics can not only inhibit soil erosion in the early stage of planting, but also improve the survival rate of vegetation. At the same time, the organic combination of artificial vegetation and grazing prohibition can effectively restore vegetation, improve vegetation coverage, and improve grassland quality.

Desertification of land and desertification of desert steppe: Using remote sensing image interpretation and ground monitoring data of meteorological factors, analyzing the distribution, dynamic change process and characteristics of aeolian desertification land; proposing population, community, ecosystem, and other ecological principles, the self-sustaining mechanism of community plants in alpine and arid sandy land was studied. In order to provide scientific basis and practical technology for regional ecological restoration and environmental comprehensive improvement, comprehensive matching technology of desertification land control, management and breeding technology of good varieties and strong seedlings of suitable plants were putting forward.

Forage production and spring grazing rest technology: High-quality hay is the main forage source during the rest of grazing, which is the basic guarantee for spring grazing rest. A series of agronomic measures such as dogmatization, plowing, leveling, fertilization, and sowing are adopted to build an annual artificial grassland on the degraded secondary bare black soil beach. The livestock were kept in captivity with the forage material produced, and the pasture was rested for 2 months until the livestock were transferred from winter and spring pasture to summer pasture.

Artificial Agricultural Improvement and Construction of Artificial Grassland: It is common to use agricultural measures to artificially promote the restoration of degraded grassland. It can achieve good benefits by loosen soil, light rake, shallow plowing, and sowing. Improving the production capacity of artificial forage is a strong support for the restoration and reconstruction of degraded grassland. Overgrazing and livestock imbalance are the main causes of grassland degradation (Liu et al. 2002). Too much grass and more livestock and serious imbalance are the focus of contradiction. Only by expanding the production capacity of artificial forage and strengthening the material basis of livestock production can be possible to improve the individual production performance of livestock, speed up the turnover of livestock, realize "returning grazing to grassland," maintain good health, and promote the transformation of grassland animal husbandry from traditional extensive management to intensive and semi-intensive management.

7.3   Sustainable Use of Rangeland Resource

Grassland is not only an important economic value, but also an important base for animal husbandry production and regional economic development. It also has important ecological service value. It has important ecological environment in climate regulation, wind and sand control, soil and water conservation, water conservation, air purification, carbon absorption, and biodiversity protection. Sustainable utilization of grassland resources refers to the management mode which can maintain the long-term and sustainable utilization of grassland, maintain the benign circle of grassland environment, and do not damage the demand of grassland resources for future generations of mankind (Chen et al. 2016a, b).

7.3.1   Grassland Improvement and Artificial Grass Planting in Pasturing Areas

The number of livestock in pasturing areas of China has exceeded the natural grassland carrying capacity. Meanwhile, the increase of grassland desertification area has aggravated the pressure of grassland grazing. Therefore, only improving grassland and planting grassland artificially can alleviate the pressure of overgrazing and improve the grassland productivity and the livestock products continuously (Chen et al. 2017).

Grassland improvement measures (e.g., sand–alkaline biotechnology can reduce soil salinity and alkalinity) can improve soil physical structure and chemical properties; fertilization can supplement the loss of mineral elements in the grassland to increase soil fertility; fencing can prohibit grazing in the growing season and promote grass production. Vegetation growth resumed. According to the different stages of vegetation retrograde succession and the soil characteristics and water conditions of different grassland types, suitable plant species can be selected according to local conditions in the process of artificial grassland planting, so that the grassland vegetation can be a resuccess and the grassland coverage and carbon sequestration potential can be increased, such as the moderately degraded grassland in Inner Mongolia and Gansu Province. On the other hand, if the soil has become semi-mobile–mobile dune, then planting sand canopy, small yellow willow, and so on can adapt to its environment.

Only when the soil becomes a fixed sand dune can we plant wheatgrass, Astragalus membranaceus and Lespedeza. After grassland improvement and artificial grass planting, grassland coverage and forage productivity were improved, and sustainable utilization of grassland resources and development of animal husbandry were promoted.

7.3.2 Strengthening Laws and Regulations and Implementing Grassland Law

Implementing grassland law is an important way to legally manage grassland and an important guarantee for sustainable utilization of grassland resources. Since the first prairie law of the People's Republic of China was enacted in 1985, China has formulated a series of regulations and suggestions on grassland protection. In 2003, China revised the prairie law further. Although there are laws to follow, the law enforcement needs to be improved everywhere (Liu et al. 2002).

Therefore, the government should strengthen grassland law enforcement and crack down on grassland reclamation, random mining, excessive grazing, and other acts. At the same time, it should also emancipate our thinking, consider involving enterprises and non-governmental organizations, and encourage the establishment of "grassland cultivation and protection companies" and "grassland protection associations." Encouraging grassland protection agencies with "economic subsidy" and "policy exemption" to achieve grassland ecological protection is not only by the government, but also a three-dimensional project of grassland ecological protection.

7.3.3 Control the Excessive Utilization of Grassland

In history, the grassland property rights system was accumulated by nomads in the long-term experience of animal husbandry production and life. Today's prairie property rights evolved from the prairie property rights in history. In the traditional nomadic period, the system of grassland common use or tribal use was practiced.

The definition of grassland property rights was not clear, and herdsmen could be nomadic on a large scale. Before liberation, grasslands and livestock were privately owned by livestock, while grasslands and livestock were collectively owned during the period of collectivization.

Therefore, in the period of collectivization or earlier, grassland property rights have reasonable exclusiveness; grazing has a certain space for movement and can be carried out in a large range of nomadic and rotational grazing, while the population and livestock are less, grazing pressure is less, there is no grassland degradation and public land tragedy, and there is no waste of public resources. Since the reform and opening-up, in 1981, the state has implemented the "grassland public ownership, contractual management" approach, that is, "grassland and livestock double con-tractual" responsibility system, grassland and livestock distribution to households, and grassland boundaries are clear. However, in the early stage of the implementa-tion of the double contractual responsibility system for grass and livestock, due to the unclear grassland boundaries and property rights, and the pursuit of short-term economic interests of herdsmen, excessive overgrazing led to serious degradation of grassland, resulting in the tragedy of public land.

In 1997, the government promulgated a policy to implement the household con-tract responsibility system, making the demarcation of grassland clear. The demar-cation of grassland is divided into households, and a 30-year contract system is implemented. Herdsmen's net fencing terminates the common property rights and moving grazing, which limits the four-season grazing and two-season grazing. In addition, excessive grazing and serious degradation appear in some grassland. At the same time, some grassland was idle and wasted. From the perspective of the evolution of grassland property rights system, the strict definition of grassland prop-erty rights boundaries is conducive to raising herdsmen's awareness of grassland protection and avoiding Tragedy of the Commons; however, the implementation of grazing space and rotational grazing is limited by the fencing of grassland to a certain extent. Therefore, grassland boundaries should be reasonably divided on the basis of grassland growth conditions, livestock quantity, and population size, achieving the goal on scientifically managing and rationally utilizing grassland resources, so as to achieve the balance between grassland and livestock, and promote the sustainable development of grassland.

7.3.4 Improving the Compensation Mechanism for Grassland Ecological Construction and Strengthening the Reward Mechanism for Ecological Protection

Compensation for grassland ecological construction is paid by the beneficiaries, that is, the government or specific beneficiaries should make necessary compensation for those who pay for the implementation of ecological environment protection, in order to enhance the enthusiasm of ecological environment protectors and builders.

It is the key research points for the compensation mechanism of grassland ecological construction to construct proposition of sustainable and effective grassland ecological and build a well-off society in an all-round way. In 2007, the No. 1 document of the Communist Party of China put forward to "to explore the establishment of grassland ecological compensation mechanism" and strengthen the grassland ecological compensation mechanism research. However, there are some problems in the implementation of compensation policy for grassland construction, such as lack of unified norms and technical standards, lagging infrastructure construction, single compensation standard, short in compensation period, lack of supporting funds, lack of modern science and technology for farmers and herdsmen.

Therefore, it is necessary to perfect the legal mechanism of compensation for grassland ecological construction, determine the compensation standard and time limit rationally, establish a long-term mechanism for grassland ecological compensation, and promote the rational and sustainable development of grassland ecological construction. On October 12, 2010, the State Council decided to establish a grant and reward mechanism for grassland ecological protection. Since 2011, the grant and reward mechanism for grassland ecological protection has been established in eight major grassland pastoral areas, including Inner Mongolia, Xinjiang (including Xinjiang Production and Construction Corps), Tibet, Qinghai, Sichuan, Gansu, Ningxia, and Yunnan. The mechanism includes the implementation of forbidden grazing subsidies, the balanced incentives for grass and livestock, the implementation of productive subsidies for herdsmen, the strengthening of support for education and training of herdsmen in pastoral areas, and the promotion of employment transfer and sustained income growth of herdsmen. Subsidies and support will be given to herdsmen in production, life, education, science and technology, so as to promote grassland ecological protection, increase herdsmen's income and sustainable development of grassland.

7.3.5 Establishing a Comprehensive System of Grassland and Livestock Products Industrialization According to Local Conditions

Because of the large-scale grassland degradation in China, it is very difficult to recover comprehensively in a short term. Therefore, it should adhere to the principle of combining prevention and control, select suitable areas for intensive grassland management relying on local conditions, and promote grassland comprehensive restoration and management by introducing funds and technology. Artificial and semi-artificial grassland are vital forms of intensive grassland production (Xu et al. 2017). The area suitable for intensive grassland production in China is 400–670,000 km^2, according to preliminary investigation, in areas with annual precipitation above 300 mm and suitable for planting grassland. It is difficult to construct animal husbandry market system because of the highly dispersed and fluidity of animal husbandry production in pastoral areas.

Therefore, it must give full play to the non-market organizational forms, such as the implementation of industrial management, the combination of leading enterprises and herdsmen to overcome the limitations of animal husbandry production and management in pastoral areas. Pastoral areas can rely on their own resources to implement the integration of production, supply and marketing, trade, industry, and animal husbandry. Speeding up the opening-up of the economy to the outside world, absorb enterprises, funds, and technology from home and abroad extensively, promote the comprehensive development of grassland animal husbandry by industrialized management, and form a comprehensive system for the industrialized development of grassland and animal husbandry.

7.3.6 Adjusting Ecological Restoration Project Reasonably

Since 1999, the Chinese government has implemented a series of grassland ecological environmental protection projects, including Green for Grain Project, Returning Grazing to Grassland Project, Wind and Sand Control project in Beijing Tianjin Tangshan, and the project of constructing grassland nature reserve. Practice shows that the implementation of these projects has achieved good ecological benefits. However, there are still deficiencies in the implementation of the project.

Firstly, the implementation areas of the project are mainly concentrated in the grassland distribution areas with severe degradation and extreme degradation, while the grassland with no degradation, mild degradation, and moderate degradation is not paid enough attention. The neglected areas will show different degrees of degradation under adverse climatic conditions and human activities. This is the reason for the partial improvement and overall deterioration of ecological restoration efficiency.

At present, grassland ecological restoration projects, especially in Horqin Sandy Land, Maowusu Sandy Land, Shaanxi Yulin, and other places, have been implemented with great efforts. Especially in Inner Mongolia, the trend of artificial grass planting area, fencing area, and forbidden grazing area directly dominated the overall change of the above indicators. However, the implementation measures of ecological restoration in Tibet and Xinjiang are relatively small. For example, in Inner Mongolia, human activities play a leading role in grassland restoration, while in Tibet and Xinjiang, the contribution of human activities is small. Therefore, the implementation of ecological restoration measures should be strengthened, not limited to severely degraded grasslands and desertification areas.

In the process of vegetation restoration, the species selection of artificial grassland is relatively simple and cannot be adjusted according to local conditions, resulting in a low survival rate of artificial grassland, and even due to lack of water caused by grassland degradation in arid and semi-arid areas. The implementation of the project lacks sustainability, for example, part of the grassland has been reclaimed under the circumstances of reduced government investment and weak supervision.

In view of the above problems, the state should make reasonable adjustments to the implementation of the ecological restoration project, further increase financial input, strengthen supervision, so that the implementation of the project can continue to play a role, while strengthening the rationality and sustainable development of the grassland restoration project.

7.4 Construction of Grassland Green Ecological Protective Screen

Constructing green ecological protective screen in pastoral areas is to restore the functions of basic ecological processes, such as plant species in richness and multiscale, vertical structure of vegetation and soil, horizontal pattern of grassland ecosystem components, heterogeneity of components of grassland ecosystem, and water–energy–material flow. Specifically, there are generally four aspects: (1) restoring degraded habitats, (2) increasing productivity on degraded land, (3) removing interference in the controlled area, and (4) making rational use and protection of the existing grassland ecosystem and maintaining its service function.

In the process of building green ecological protective screen in different areas, levels of construction requirements should be formulated differently according to different social, economic, cultural, and living needs: (1) realizing the stability of the surface basement of the ecosystem. It is difficult to ensure the sustainable succession and development of grassland ecosystem; (2) restoring vegetation coverage and soil fertility; (3) increasing plant species composition and biodiversity; (4) achieving the recovery of the biological community, improving the productivity and self-sustaining ability of the ecosystem; (5) reduce or control interference. Due to the regional differences of degraded grassland ecosystems, as well as the different types and intensities of external disturbances, the degradation types, stages, processes, and response mechanisms of the ecosystems are different too. Therefore, the restoration objectives, emphasis, and the key technologies selected are often different in the process of restoration and management of different types of degraded grassland ecosystems. Nevertheless, for degraded ecosystems generally, the following basic restoration technologies are involved (Meng and Gao 2002): (1) restoration technologies for abiotic or environmental factors (including soil, water and atmosphere), (2) restoration techniques of biological factors (including species, populations, and communities), and (3) overall planning, design, and assembly technology of ecosystem (including structure and function).

Since 2003, a series of green ecological protective screen protection and construction projects have been deployed in the three-river source region—Qinghai Lake Basin, Tibet Plateau, and Qinghai–Tibet Plateau. Luo et al. (2009) carried out the ecological effect evaluation of the first-phase project of ecological protection and construction of three-river source region, and reached the conclusion that the trend of ecosystem degradation has been curbed preliminarily, the ecological situation of the key ecological construction projects has improved, and the long-term and arduous task of ecological construction has become prominent (Xu et al. 2017).

The unique ecological protective screen function of Tibet Plateau is manifested in its water conservation and hydrological regulation to China and even East Asia, reducing the impact of plateau dust on surrounding areas and providing habitats for plateau-specific biodiversity (Huang et al. 2018). The ecological protective screen in Tibet is threatened by the climatic background of marked increase in temperature, precipitation, sunshine hours, and snow days, the function, and human activities. Long-term and continuous ecological compensation and incentive measures should be implemented to ensure the effective protection of fine ecological resources by reducing ecological pressure in ecological protective screen function. On the basis of the scientific evaluation of the achievements in the ecological projects already carried out, and in accordance with the needs of long-term management and consolidation of the achievements, we will continue to invest funds and establish a long-term mechanism for the consolidation of targeted achievements.

In the process of building Tibet's ecological protective screen, the protection and construction project of Tibet's ecological protective screen protective screen was initiated and implemented in 2008, with near-term planning to 2015 and long-term planning to 2030, the focus and core of the project is grassland ecological protection and construction in freeze–thaw areas, including protection, construction, and support of ten projects of three categories. In 2015, the contradiction between grassland and livestock in Tibet has been obviously alleviated, the control of sandy land and soil erosion in key areas has been significantly promoted, and the progress of biodiversity protection in the headwaters of major rivers, lakes, wetlands, and valleys has been accelerated. The ecological environment monitoring system and monitoring network have adapted to the demand of ecological environment construction and protection basically. To be sure, the role of traditional energy substitution in supporting the protection of ecological environment has been significantly strengthened, and the ecological environment and economic society have begun to step into a coordinated development track (Gao et al. 2010; Wang and Wu 2013).

Grassland ecological protection, the monitoring and evaluation system of grassland ecological protection in Inner Mongolia has been implemented in an all-round way, the implementation plan has been formulated scientifically, the ground survey has been organized conscientiously, and the remote sensing technology has been applied to the analysis. By 2016, the cold-season-edible herbage reserves and suitable livestock carrying capacity of 33 natural grasslands in animal husbandry banners have been predicted (Dai et al. 2016). The grassland supervision and Administration Bureau and the grassland survey and Planning Institute of Inner Mongolia used ground monitoring and satellite remote sensing technology, combining with meteorological data, analyzed the cold season forage reserves of 33 natural grasslands in the region, and calculated the cold season edible forage reserves and suitable stocking capacity of natural grasslands in 2016. The natural grassland of 33 animal husbandry banners has 7.56 billion kg of edible herbage and 18.299 million units of sheep in cold season. Affected by drought in some areas, they have decreased by 16.4% and 20.6%, respectively.

Xinjiang has made unremitting efforts in curbing and harnessing the "three aspects of grassland" and improving the grassland ecological environment. Especially since entering the new century, the state regards the restoration and construction of the ecological environment in western region as an important measure to implement the strategy of developing the western region. Xinjiang has successfully implemented a large number of grassland ecological construction projects and achieved remarkable results, such as natural grassland restoration and construction projects, natural grassland fence construction projects, natural grassland grazing project, grassland disaster prevention and control. Up to 2012, a total of 189 million mu of natural grassland had been fenced, accounting for 26% of the available natural grassland in Xinjiang. Among them, 3.23 million ha was forbidden, 0.64 million ha was rested, 16.2 million mu was rotational grazing in zoned areas, and 1.298 million mu was newly constructed with artificial forage land. Gazing and returning grassland project continuously has played a great role in promoting the restoration of natural grassland (Yu et al. 2016).

The grassland supervision center of the Ministry of agriculture has no doubt about Xinjiang's natural grassland returning to pasture and grassland project area for many years, and the grassland vegetation coverage, height, and aboveground biomass in the temperate desert project area were increased by 5–10%, 1–4%, and 5–14%, respectively. The average coverage of grassland vegetation increased by 10%, the average height increased by 2.6 cm, and the average aboveground biomass increased by 17.6, in mountain meadow engineering area (Yu et al. 2016). The average coverage of grassland vegetation increased by 5–15 percentage points, the average height increased by 3–16 cm, and the average aboveground biomass increased by 10–24%, in lowland Meadow engineering area. The overall deterioration of grasslands has been curbed, and local ecological condition was improved (Yu et al. 2016).

References

Akiyama T, Kawamura K (2007) Grassland degradation in China: methods of monitoring, management and restoration. Grassl Sci 53(1):1–17

Chen JM, Pavlic G, Brown L, Cihlar J, Leblanc SG, White HP (2002) Derivation and validation of Canada-wide coarse-resolution leaf area index maps using high-resolution satellite imagery and ground measurements. Remote Sens Environ 80(1):165–184

Chen YZ, Mu SJ, Sun ZG, Gang CC, Li JL, Padarian J (2016) Grassland carbon sequestration ability in China: a new perspective from terrestrial aridity zones. Rangel Ecol Manag 69(1):84–94

Chen YZ, Ju WM, Groisman P, Li JL, Propastin P, Xu X (2017) Quantitative assessment of carbon sequestration reduction induced by disturbances in temperate Eurasian steppe. Environ Res Lett 12(11):115005

Dai E, Huang Y, Wu Z, Zhao D (2016) Spatial-temporal features of carbon source-sink and its relationship with climate factors in Inner Mongolia grassland ecosystem. Acta Geogr Sin 71(1):21–34

Gao QZ, Li Y, Wan YF, Jiangcun WZ, Qin XB, Wang BS (2010) Significant achievements in protection and restoration of alpine grassland ecosystem in northern Tibet, China. Restor Ecol 17(3):320–323

Gao QZ, Wan YF, Li Y, Guo YQ, Ganjurjav H (2013) Effects of topography and human activity on the net primary productivity (NPP) of alpine grassland in northern Tibet from 1981 to 2004. Int J Remote Sens 34(6):2057–2069

Huang L, Cao W, Xu XL, Fan JW, Wang JB (2018) The ecological effects of ecological security barrier protection and construction project in Tibet Plateau. J Nat Resour 33(03):398–411

Jia HT (2007) Ecological effects of enclosure on degraded grassland in Xinjiang. Xinjiang Agricultural University

Li B (1997) The degradation of grassland in North China and its countermeasure. Sci Agric Sin 30:1–10

Liu ZL, Wang W, Hao DY, Liang CZ (2002) Prebes on the degeneration and recovery succession mechanisms of Inner Mongolia Steppe. J Arid Resour Environ 16(1):84–91

Liu L, Zhang Y, Bai W, Yan J, Ding M, Shen Z (2006) Characteristics of grassland degradation and driving forces in the source region of the Yellow River from 1985 to 2000. J Geogr Sci 16(2):131–142

Luo L, Wang ZM, Song KS, Zhang B, Liu DW, Ren CY (2009) Research on the correlation between NDVI and climatic factors of different vegetations in the Northeast China. Acta Bot Boreali-Occident Sin 4:800–808

Men L, Gao HW (2002) The situation, causes and rehabilitation of degraded grassland in China. In: China international grass industry development conference and the Sixth Congress of China Grassland Society, Beijing, China, p 4

Ren H, Shen WJ, Lu HF, Wen XY, Jian SG (2007) Degraded ecosystems in China: status, causes, and restoration efforts. Landsc Ecol Eng 3(1):1–13

Rigge M, Wylie B, Zhang L, Boyte SP (2013) Influence of management and precipitation on carbon fluxes in great plains grasslands. Ecol Indic 34:590–599

Sun Q, Li B, Zhang T, Yuan Y, Gao X, Ge J (2017) An improved Biome-BGC model for estimating net primary productivity of alpine meadow on the Qinghai-Tibet Plateau. Ecol Model 350:55–68

Wang JF, Wu QB (2013) Annual soil CO_2 efflux in a wet meadow during active layer freeze-thaw changes on the Qinghai-Tibet Plateau. Environ Earth Sci 69(3):855–862

Wen F, Ping Z, Bo C, Wenyan Z (2007) Some scientific problems of grassland degradation in arid and semi-arid regions in Northern China. Chin J Grassl 29(5):95–101

Xu X, Yang G, Tan Y, Tang X, Jiang H, Sun X (2017) Impacts of land use changes on net ecosystem production in the Taihu Lake Basin of China from 1985 to 2010. J Geophys Res Biogeosci 122(3). https://doi.org/10.1002/2016jg003444

Yan YE, Wang JH, Shi JZ, Zhou XL, Xiao-Zhou WU, Zi-Jun LV (2010) Analysis on grassland resources and their deterioration situation on North Slope of Qilian Mountains. Pratacultural Sci 27(7):24–29

Yu H, Xun QL, Zhang YH, Zhang QQ, Zhu GH, An SZ (2016) Discussion on grassland ecological function regionalization in Xinjiang. China Anim Husb Vet Med 43(4):1118–1124

Zhang WH, Yang W (2011) The feature analysis for grassland degradation and the restoration of natural vegetation in degraded grassland. North Environ 8:40–44

Zhang W, Zhang H, Ze B (2006) Progress studies on the carbon cycle of Alpine meadow in China. J Mt Sci 24(B10):266–274